T0291831

WAVES

The MIT Press Essential Knowledge Series

WAVES

FREDRIC RAICHLEN

The MIT Press | Cambridge, Massachusetts | London, England

© 2013 Fredric Raichlen

All rights reserved. No part of this book may be reproduced in any form by any electronic or mechanical means (including photocopying, recording, or information storage and retrieval) without permission in writing from the publisher.

Set in Chaparral by the MIT Press.

Library of Congress Cataloging-in-Publication Data

Raichlen, Fredric.
Waves / Fredric Raichlen.
    p.   m.—(The MIT Press essential knowledge)
Includes bibliographical references and index.
ISBN 978-0-262-51823-9 (pbk. : alk. paper)
1. Water waves. 2. Ocean waves. 3. Wave mechanics. I. Title.
TC172.R35 2012
551.46'3—dc23'

                                                    2012015335

This book is dedicated to my wonderful grandchildren:
Cody, Jacob, and those to come.

CONTENTS

CONTENTS

SERIES FOREWORD

The Essential Knowledge series offers accessible, concise, beautifully produced pocket-size books, written by leading thinkers, on topics of current interest in a variety of fields, ranging from the cultural and the historical to the scientific and the technical.

In the information age, opinions, rationalizations, and superficial descriptions are readily available. Much harder to come by are the foundational knowledge and the principled understanding that should inform our opinions and decisions. Essential Knowledge books fill those needs. They synthesize important subject matter to help readers navigate a complex world.

Bruce Tidor
Professor of Biological Engineering and Computer Science
Massachusetts Institute of Technology

PREFACE

Anyone who has spent a day at a beach or on a boat probably has wondered at the power and mechanics of the ocean. Years ago, as I sat on a beach with my two sons, they asked me many questions about the waves offshore. Where do waves come from? How do they form? What happens as they approach the shore? How do they break? Why do they break? Like the waves, the questions kept coming.

As a coastal engineer who has spent more than fifty years studying waves and their coastal effects, I believe I can provide answers to these questions. In this book, I attempt to describe the complex phenomena of water waves and their coastal effects in a simple way.

In recent years, we have witnessed disastrous tsunamis in Sumatra (2004) and in Japan (2011). After the 2010 oil spill in the Gulf of Mexico, we held our collective breath waiting to see what would wash ashore and how it would impact the environment. Hurricane Katrina is another example of the potentially destructive power of the coastal waters. All of us should increase our understanding of water waves, both the everyday ones and the exceptional ones. This book is a start.

Aside from ocean waves, many other matters having to do with the coasts and shorelines are of interest. For example, when looking seaward from a beach you may see something that appears to be a pile of rocks extending out from the shore. What is that odd structure meant to protect? How was its placement determined? Questions about the beach itself may come up, too. Why is it we have a wonderful beach to sprawl on at some times of the year, but at other times there is essentially no beach at all at the same location? These are just a few of the question I'll take up.

Some portions of the book are more technical than others. A reader can skip some sections altogether and still get a good understanding of the basics of coastal engineering. There are very few equations, and those that are presented are presented simply, and also in plain English. (They are presented in more "technical" style in an appendix.)

Whatever your background, this book will answer the first wave of questions about the coastline and will give you a greater understanding of the wave-related phenomena you may observe the next time you go to the beach.

ACKNOWLEDGMENTS

The idea for this book came from my son David, who provided assistance during its preparation. Both he and my other son, Robert, have lent a sympathetic ear throughout the course of the writing. They have both encouraged me along with Robert's wife Amy and David's fiancée Sarah. Steven Raichlen, my nephew, has been a sounding board on many different matters. A major motivation for this effort has been to provide answers to some of these wave questions for my grandsons Cody and Jacob. At the MIT Press, Marguerite Avery and Katie Persons were extremely helpful in many ways, both before and during the book's preparation, and the editing was very helpful. Last but certainly not least, my wife Judy has encouraged and supported this effort and me throughout my career at Caltech. Without her I would not have gotten to the point where this book could be written.

ACKNOWLEDGMENTS

WAVES

Why should we be interested in water waves? Other than curiosity, our primary interests are in how they affect the shore and how we can protect ourselves from injury, loss of life, and property damage. For these reasons, the questions addressed in this book relate to both the science of water waves and the engineering problem of how we cope with what nature throws at us.

Many people around the world live near a coast and use the near-shore waters for recreation and for sustenance. Beaches are important assets to countries worldwide. Anyone who lives near a coast has seen how major storms affect beaches. And the operation of a port is directly affected by normally occurring ocean waves as well as by extreme events such as hurricanes and tsunamis. The waves we enjoy for recreation may cause the interruption of shipping and disrupt the flow of goods around the world.

Events of the past few decades show how important it is to understand our ocean environment. Hurricanes Camille (1969), Hugo (1989), and Katrina (2005) caused the loss of many lives and billions of dollars in damage in the United States alone. And in recent years the destructive power of tsunamis has been evident. Tsunamis, much like earthquakes, are natural phenomena that can't be predicted, but we must be vigilant about them and learn how to protect ourselves from their consequences.

What Is a Wave?

Regular Waves—Basic Definitions

A water wave is shown schematically in figure 1. A group of these identical sinusoidal waves are denoted as a *regular wave train*. A wave's *amplitude* is the distance from the undisturbed ocean surface to the displaced water surface. A wave's *crest* is the point of maximum elevation of the water surface of the wave, and the distance from the undisturbed ocean surface to that point is referred to as the *amplitude of the crest*, conventionally denoted as a_c. The depression of the water surface below the undisturbed ocean surface is called the *trough*; its minimum is denoted a_t. *Wave height* (H) refers to the distance between this minimum and the maximum elevation; *wave length* (L) is the distance from one position on the wave to the following like position in a group of waves (for example, from one wave crest to the following crest). If you are at one location and these waves pass you, the time between like positions on the wave is called the wave's *period* (denoted by T). If the waves are uniform, such as those shown in figure 1, both the wave's length and its period are constant.

The speed with which the waves would seem to travel past an observer standing at that fixed position is referred to as the *wave celerity* or the *phase speed* or simply the *speed* and is denoted by C. A wave's celerity is equal to its length divided by its period (L/T).

A *small-amplitude* wave is defined as a wave whose dynamics can be defined by the simplest mathematics. If more complex mathematics is needed to describe a wave, it is usually called a *finite-amplitude* wave.

Deep-Water and Shallow-Water Waves

For most waves in the deep ocean, the length is directly related to the square of the period. It is useful here to recall the definition of a wave's period. If you were to keep your attention on one spot in the ocean, the period would be the time it takes for a crest (or any other position on a wave) to recur at that location. Thus, you see one wave; then, one period later, you see another. If you were to freeze the surface of the ocean and observe it from above, a wave's length would be defined simply as the distance between crests (or any other like positions on the wave).

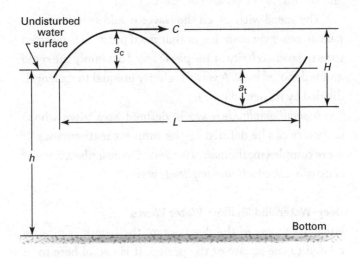

Figure 1 A definition sketch of a small-amplitude wave. Here a_c is the amplitude of the crest, a_t is the amplitude of the trough, H is the wave's height, L is the wave's length, C is the wave's speed (celerity), and h is the depth of the water.

If you consider a simply shaped wave with a sinusoidal profile (as in figure 1), you can determine its length in the deep ocean from its period.

A simple equation allows you to determine the length of a wave in the deep ocean. At the shore, you can use it to impress a companion by asking him, or her, to determine the time between crests, then scratching your head for a minute before specifying the length in the deep ocean. In deep water a wave's length is equal to 1.56 times the square of its period if the length is in meters, or 5.12 times the square of the wave's period if the length is in feet. Thus, if a wave's period is 10 seconds, its length in deep water will be 156 meters (512 feet). But what is considered deep water? That doesn't depend on the absolute value of the depth; it is relative. If the ratio of the depth to the length is greater than 1/2, a wave is defined as a deep-water wave. Under those conditions, a wave doesn't "feel" the bottom. In other words, the depth could be greater and the length determined as specified above wouldn't be affected. Since the wave speed, C, is equal to the wave length divided by the wave period, for a deep-water wave the speed is equal to 1.56 times its period if the length is in meters, or 5.12 times the wave's period if the length is in feet.

Only if the velocity of the water particles under the wave are essentially zero before the bottom is reached does the wave not feel the effect of the bottom. Under small-amplitude waves the water particles travel in a closed elliptical path. Suppose you are floating in the ocean as a wave passes. You know from experience that you simply bob up and down and back and forth as the wave goes by. That motion is caused by the movement of the water particles beneath the wave. It is only when the wave becomes large—a finite-amplitude wave—that you don't return to where you started.

As the depth becomes greater than about twice the wave's length, the ellipses become circles. The diameter of these circles, which describe the path of water particles beneath the water surface, decrease exponentially downward from the surface, with the decrease determined by the ratio of the depth to the length. This is illustrated in figure 2. Thus, if the velocity is very small at the bottom, the wave essentially doesn't "feel" it. Hence, the bottom can vary in any wild manner and the characteristics of the wave will not change. (This is the big advantage of traveling by submarine in the deep ocean. A monstrous storm could be occurring with huge waves on the ocean surface, but at some depth below the surface the submarine just sails quietly along, not disturbed by what is happening above.)

What about the magnitude of the velocity of the water particles? It is much smaller than the wave's speed (celerity). In fact, for a deep-water wave the ratio of the *maximum* horizontal water-particle velocity (the horizontal

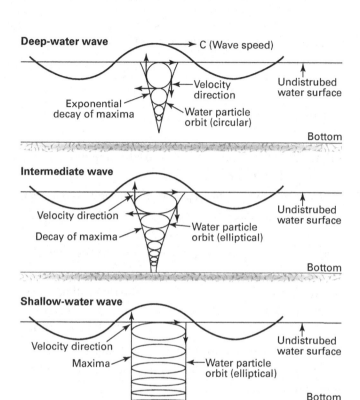

Figure 2 A schematic drawing of the orbital motion of water particles under a deep-water wave, an intermediate wave, and a shallow-water (long) wave.

water-particle velocity at the surface) to the wave's speed is equal to

$$2\pi \left(\frac{a_c}{L_o} \right)$$

—that is, 6.28 (2π) times the ratio of the amplitude of the wave's crest to its length.

Consider a wave with a period of 8 seconds and a crest amplitude of 1 meter (3.28 feet). The wave's length in deep water will be 99.8 meters (about 328 feet). Thus, the depth has to be about 50 meters (656 feet) for the water-particle velocity to be essentially zero at the bottom. The phase speed (that is, the speed of the wave shape) will be equal to the ratio of the wave's length to its period—in this case, about 12.5 meters (40.9 feet) per second. The ratio of the maximum horizontal water-particle velocity to the wave celerity (or the speed of travel of the wave shape) will be about 0.063.

Does this idea of the velocity of the water particles' being small relative to the wave's speed sound strange? Much the same is true of the speed of sound in the air: the velocity of the air particles is much less than the velocity of the sound waves. At sea level, the velocity of sound in air is about 335 meters (1,100 feet) per second. If the velocity of the air particles were of that magnitude, you would be picking yourself up off of the ground every time someone spoke. Seating wouldn't be necessary at rock concerts, because everyone would be knocked flat on the ground by the first sounds from the amplifiers.

The other extreme is when a wave's length is large relative to its depth. When a wave's length is greater than about 20 times its depth, it is called a *shallow-water wave* (or a *long wave*). (Again, let's consider only "small-amplitude" waves.) Like deep-water waves, shallow-water waves are a simplification that allows for some basic observations. For deep-water waves, the length was directly proportional to the square of the period. The length of a shallow-water wave is directly related to the period.

A closer look reveals that the speed (celerity) of a shallow-water wave is equal to

$$\sqrt{gh}$$

—that is, the square root of the product of the acceleration of gravity and the depth of the water.

What is the meaning of the acceleration of gravity? If you were to drop an object in a vacuum, it would fall freely, affected only by the gravitational attraction of the Earth. While falling, it would accelerate at a rate of 9.8 meters (32.2 feet) per second per second). That means that for every second the object falls its speed increases by 9.8 meters (32.2 feet) per second. Thus, in 2 seconds the velocity of the falling object goes from zero to 19.6 meters (64.4

feet) per second. The acceleration of gravity is usually denoted by g.

Since the speed of any wave is simply the wave's length
divided by its period, we see that the length of a shallow-
water wave is equal to the product of its speed and its
period. Thus, for both deep-water waves and shallow-
water waves the length is proportional to a function of the
period. The length of a deep-water wave is proportional to
the square of the period, whereas that of a shallow-water
wave is proportional to the period.

Under a shallow-water wave the water particles also
move in closed elliptical orbits. In that case the horizontal major semi-axis of the ellipse is constant with depth,
but the vertical minor semi-axis of the ellipse decreases
linearly with depth and becomes zero at the bottom. (The
major semi-axis is simply one half of the length of the ellipse and the minor semi-axis is one half of the width of
the ellipse.) This variation is shown in figure 2. By linearly
decreasing with depth we mean that the vertical width of
the ellipse decreases like a straight line going from a maximum at the surface to zero at the bottom. Thus, at the bottom the water particles simply move back and forth as the
wave passes over and the water particles do not move vertically. The water particle orbits under a wave intermediate
between a deep water and a shallow water wave are also
shown in figure 2. In this case the water particles move in
an elliptic orbit with the major axis decreasing with depth

but not becoming zero. For a shallow-water wave the ratio of the *maximum* horizontal water-particle velocity to the wave's speed is equal to

$$\frac{a_c}{h},$$

the ratio of the amplitude of the wave's crest to the depth.

To complete this discussion of wave dynamics, before discussing the transformation of waves as they travel, it is important to discuss the pressure under waves. The profile of a wave (perhaps modified) would register on a pressure gauge installed on the bottom of the ocean as the wave traveled over. This is the principle of how the DART (Deep-Ocean Assessment and Reporting of Tsunami) buoy developed by the Pacific Marine Environmental Laboratory (a laboratory of the National Oceanic and Atmospheric Administration) works in determining the height of a tsunami. The pressure fluctuation caused by a tsunami passing over the pressure sensor is directly proportional to the amplitude of the wave at the surface. (A tsunami is a shallow-water wave even in the deepest ocean.)

How does pressure vary beneath the ocean's surface without waves? At a given distance below the surface (with the air pressure at the surface neglected), the pressure is equal to the weight of the column of water between the location and the surface divided by the cross-section area of the water column. The pressure increases in a linear manner as you move downward. If you were to plot the pressure as a function of depth (neglecting the atmospheric pressure at the surface), you would see that it is triangular in shape, going from zero at the surface to the depth times the weight per unit volume of water at a location beneath the surface. Suppose the depth of water in a fresh-water lake is 100 meters (328 feet). The pressure at the surface, with the pressure of the air acting on the surface neglected, will be zero. At the bottom the pressure will be about 981,000 newtons per square meter, or 20,467 pounds per square foot. This linear variation of pressure with depth is called a *hydrostatic pressure distribution*.

Now add waves to the undisturbed ocean and consider what happens to the bottom pressure as the crest of a wave travels over a location. On the bottom the pressure increases compared to the pressure if there were no waves. Conversely the pressure decreases as the trough passes the location. However, for a wave with a ratio of depth to length greater than 1/20 (5 percent), because of the accelerations and decelerations of the water particles under the wave, the pressure on the bottom as expressed as the height of a water column will not be equal to the sum of the wave's amplitude and the water's depth. Beneath the crest of the wave, the maximum pressure, expressed in terms of the height of a water column, will be less than the sum of the depth and the amplitude of the wave crest. Under the wave's trough the pressure will be greater than the depth

Since water surface changes with time, at a fixed location the water particles' velocities also change with time. Therefore, there is an acceleration and a deceleration associated with a water particle when the wave passes.

minus the trough's amplitude. If the wave is a deep-water wave with a ratio of depth to length equal to 0.5, the pressure at the bottom will vary by, at most, about plus or minus 9 percent of the wave's amplitude as the wave passes.

If a pressure measuring instrument on the bottom is exposed to a shallow-water wave, the maximum pressure recorded, expressed as the weight per unit area of the height of a water column above it, would equal the wave amplitude plus the depth times the weight per unit volume of the water, and the minimum pressure would be equal to the depth minus the amplitude of the wave trough times the weight per unit volume of the water. In other words, on the bottom the pressures are not affected by vertical water particle accelerations or decelerations.

Of course, waves need not be either "deep-water waves" or "shallow-water waves." When the period of a wave varies between these two limits, the wave can be called an "intermediate-depth" wave. All of its characteristics, including its length, its speed, and the pressure under it, can be determined using the same equations that have been simplified for the two extremes—shallow-water and deep-water waves. (For a more complete mathematical treatment, see Ippen 1966 or Dean and Dalrymple 1984.)

The Generation of Ocean Waves
Waves are generated by wind in a complex way. Hermann Ludwig Ferdinand von Helmholtz (1888, 1890) proposed

that small oscillations induced at the interface between two fluids became a group of waves that traveled at speed C. Lord Kelvin (Thomson 1887) obtained the same result as Helmholtz, and found that, if the upper fluid was air and the lower fluid was water, the speed of the wave generated was the same as one obtains from linear wave theory for a deep-water wave.

In 1924, the geophysicist Harold Jeffreys observed waves forming on a pond in light winds. He concluded that waves began forming on the water surface when the wind's speed approached about 110 centimeters per second (3.6 feet per second) and the initial distance from the crest of one wave to the next was about 6 to 8 centimeters (2.4 to 3.1 inches). The waves appeared to travel at a speed of about 30 centimeters per second (about 1 foot per second). These waves are referred to as *capillary waves*, although the wave length for the minimum speed of a capillary wave is approximately 2 centimeters. Jeffreys presented a "sheltering hypothesis" according to which the pressure on the windward face of the wave exceeded that on the leeward face. He proposed that energy was transmitted from the wind to the wave as long as the speed of the wind was equal to or greater than the celerity of the waves.

The oceanographers Harald Sverdrup and Walter Munk (1947) developed a theory of the transfer of energy from wind to waves that included both normal and tangential stresses. That theory was the basis for their

The general process of wind wave generation goes from the wind ruffling the water's surface to the feeding of energy into this ruffled surface by the wind, resulting in the growth of the waves.

wave-forecasting methods. Later, Owen Phillips (1957) considered pressure and surface stress fluctuations in the earliest stages of wave generation. Klaus Hasselmann (1961, 1962) discussed the flux of wave energy in ocean waves resulting from the nonlinear interactions among different wave components.

The roughening of the ocean's surface is best understood in terms of how the wind's velocity varies from a high altitude to the sea surface. This vertical variation of the velocity is called the *velocity profile*. The velocity profile is affected by the viscous properties of the fluid. (For example, for the same velocities and distances from a surface the effects of the viscosity on the velocity distribution of air flowing would be very different from that of water or molasses flowing in the same manner.)

At a solid surface, such as that of the land, the fluid velocity must become equal to zero, since the surface is stationary. Therefore, there would be a variation of velocity from the value at a large distance above the surface to zero at the land's surface. In a similar manner, over the ocean the velocity decreases as the sea surface is approached, although it may not become zero at the surface. And when velocities are very small relative to the ratio of a characteristic length to a measure of the viscosity, the flow is very "quiet" and can be said to move in layers, even though it still decreases as the surface is approached. Such a flow is called *laminar*. In the case of a laminar flow, the velocities and pressures at a certain elevation are steady with time. However, the flow becomes unsteady as the velocity increases for the same fluid and characteristic length. That means that velocities and pressures in the flow vary with time at a given elevation. In fluid mechanics such a flow is referred to as *turbulent*. In turbulent flow the velocity and the pressure can be pictured as composed of a steady part and a fluctuating part.

It is the fluctuating parts of the pressure and shear induced at the water surface by the velocity that cause the water surface to become ruffled. The constant part of the pressure and velocity acts the same on the surface everywhere, but the fluctuating pressure and velocity act differently—in fact, randomly—at various locations. As the wind blows, energy is put into the ruffled surface and the waves begin to grow in both height and length. Indeed, a point in the waves' growth and speed may be reached at which the waves now act on the wind, and the waves reach their optimal characteristics.

To summarize, the process by which wind generates waves proceeds from the wind's ruffling the water's surface by its fluctuating components of pressure and velocity to the wind's feeding of energy into this surface and the resulting growth of the waves. The interested reader with a scientific bent is referred to the bibliography for publications with a more detailed description of this complicated process.

Major advances in both theory and observations were made in the middle of the twentieth century. During World War II, it became imperative for the U.S. military to develop methods of predicting the heights of waves near beaches so that amphibious landings could be made more safely. Observations made for this purpose by Walter Munk and his colleagues at the Scripps Institute of Oceanography in La Jolla led to a greater understanding of how wind generates waves. After the war, a number of theoretical advances were made in an attempt to understand this complex problem; a few of these were summarized above. But what the engineering profession needed was the ability to simply describe the characteristics of waves leaving the area of a storm at sea on the basis of the velocity of the wind and the size and duration of the storm.

In the 1940s, Sverdrup and Munk collected a wide range of wave measurements, primarily from shipboard observations. These measurements allowed them to predict the heights and speeds of waves from a knowledge of the wind speed, the *storm fetch* (the length of the region exposed to the winds), and how long the wind blew over the fetch. Data were added to this work in the 1950s by Charles Bretschneider, who modified the predictive curves developed by Sverdrup and Munk and came up with what is usually referred to as the SMB (Sverdrup-Munk-Bretschneider) approach. This work resulted in a set of charts that could be used to predict a representative wave

The sea surface appears as if a number of waves with different amplitudes, frequencies, directions, and phases were superimposed, resulting in a relatively random variation of the water surface elevation and the wave lengths.

height and wave speed given the variables mentioned above. (See Bretschneider 1952.)

In the mid 1950s, Willard Pierson, Gerhard Neumann, and Richard James understood that it was insufficient to talk about only an estimate of a wave's height and its speed (or its period), as the SMB approach did, and that it was more appropriate to approach wave generation from spectral considerations. Their approach is denoted as the PNJ method. (See Pierson, Neumann, and James 1955.)

About the spectrum The concept of a spectrum can perhaps be best attributed to Sir Isaac Newton, who found that sunlight could be decomposed into a range of colors by a prism. The spectrum of light shows how the light's intensity varies with the colors, each of which has a specific wavelength. This concept of considering a quantity composed of individual components, each with a specific magnitude and frequency, has been extended to many other fields. (When the spectrum is represented graphically usually the energy is plotted on the vertical axis and frequency on the horizontal axis.) Consider the spectrum of a sound. If the sound is one pure frequency, the spectrum that indicates the distribution of sound intensity (or energy) with frequency will be a vertical line at the frequency of the signal. However, if the sound has intensities that are the same for a wide range of frequencies, the energy spectrum will consist of a horizontal line at the energy corresponding to that of these frequencies. This is sometimes called a "top hat" distribution. Generally the energy distribution of speech falls between these two extremes.

The spectrum of the sea's surface is much like the spectrum of light or that of sound. The energy of the waves at a specific frequency (or wave length) is proportional to the square of the amplitude of these waves at that frequency. Thus, the spectrum of the energy distributed with frequency of a train of waves is directly related to the distribution of wave amplitudes with frequency (or wave length). When the sea surface is observed at one location, as with sound, the variation is generally not like a simple sine wave with a fixed amplitude and a fixed frequency. The sea's surface appears as if a number of waves with different amplitudes, frequencies, directions, and phases have been superimposed, resulting in a more random variation in the surface elevation and the frequency of the disturbance. In other words, both the amplitude of the surface motion and the period (or frequency) vary with time. A true sea spectrum would also have direction associated with it, since wave energy approaches a particular location from many directions. At a given location, the energy spectrum of these changes in the water surface elevations with time shows the contribution of the energy of waves at a particular frequency (or wave length) and direction to the total energy of the sea surface. Figure 3 shows examples of a spectrum of the distribution of energy with frequency

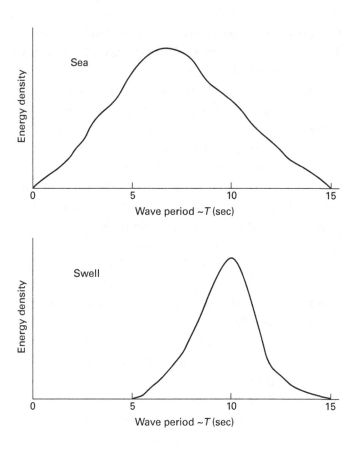

Figure 3 An example of a spectrum of the sea surface for the condition of sea and of swell. The ordinate is an estimate of the wave energy and the wave period is plotted as the abscissa.

for waves just leaving a storm region (sea) and for waves a distance from the storm (swell) for the case of waves coming only from one direction.

On rare occasions, the sea's surface can take on a much more organized appearance. A startling example of a group of large regular waves in the ocean is shown in figure 4. This aerial photograph, taken by the U.S. Army Air Corps around 1933, shows several biplanes flying over a sea with what appear to be uniform corrugations. The waves appear to have crests with very flat troughs between them. These waves, which are very different from the sinusoidal ones shown in figure 1, are referred to as *finite-amplitude waves*. They can be generated in a laboratory, but to see them in nature is surprising. A light breeze seems to have generated a cross-chop of smaller waves traveling at an angle to the larger ones. Thus, ocean waves with all of their complexities can sometimes display surprising simplicity.

An important concept of the PNJ and SMB methods for predicting the heights and the periods of waves is that it takes time and distance for waves to develop in a storm. It is probably intuitive that if a wind blows a long time over a small distance (a small fetch) the waves' heights will be less than they would have been had the same wind blown over a much longer distance (a larger fetch) for the same duration. Likewise, for a given fetch, the longer a wind blows, the larger the waves that will result. There is an upper limit to this growth with regard to fetch or duration,

OVER THE PACIFIC MAJ. GEN'S MARTIN & GRAVES HOP ALBROOK C.Z. TO DAVID, R.P.

Figure 4 An aerial photograph of large regular waves with a cross-chop taken near the Panama Canal Zone around 1933.

and when that limit is reached the waves traveling (propagating) from the storm region will be fully developed. Otherwise they will be limited in height by the fetch or by the duration of the wind. If you think about the earlier development of waves from high-frequency (short-period) pressure fluctuations and from ruffling of the ocean's surface, you quickly realize, considering the spectral concept, that the growth in the wave field starts at higher frequencies and proceeds to lower frequencies. The PNJ concept is that the spectrum develops in this manner from higher to lower frequencies until the sea is fully developed. In the 1970s, Klaus Hasselmann collected an extensive set of data in the North Sea and showed that there could be a common spectral shape for storm waves.

With the advent of large-scale computing facilities, these somewhat simplistic approaches gave way to numerical models that appear to predict the water waves reasonably well from particular wind characteristics and storm size and duration. Numerical methods are employed for purposes ranging from determining the wave environment to solving problems of coastal engineering to predicting wave conditions for surfing. For example, the Coastal Data Information Program (http://cdip.ucsd .edu) predicts wave conditions on the West Coast of the United States and provides warnings of extreme storms. However, engineering studies often require simple wave predictions that don't justify the use of complex computer models. The SMB and PNJ methods and wind predictions are straightforward ways of obtaining the basic characteristics of waves leaving a storm area by means of "back-of-envelope" calculations.

Irregular Waves

In the ocean, a wave's length and its period are not constant; there is variation, which may or may not be significant. In a very simplified sense, an observer looking down on a group of ocean waves sees, as a first approximation, what looks like an irregular corrugated surface with different distances between the peaks of the corrugations. Thus, the observer would see a range of wave lengths, and if the observer were at one location and watched these waves pass by he would observe a range of intervals of time between crests or periods.

Generally, what at first appears to be a random distribution of wave amplitude and wave length (or wave period) in an observation of a group of ocean waves results in a relatively peaked spectrum. In such a spectrum, the wave energy is concentrated primarily at one frequency, the *spectral peak*, with less energy at higher and lower frequencies.

Consider for a moment a sound that results in a peaked spectrum. Suppose that sound is produced by two very similar frequencies. The subsequent sound will consist of one tone that has a frequency equal to the average of the two frequencies and a second tone that has a frequency equal to half the difference between these two frequencies. This results in an amplitude-modulated sound. The frequency is constant, but the sound level varies slowly and periodically with time. We call the variation the *beat frequency* of the sound.

Water waves generated by wind behave in a similar way. At a given location, the water surface varies with time, with an apparent main period (or wave length) that appears to be modulated with a larger period leading to the appearance of groups of waves in a storm sea. The periods of these "wave groups" can be of importance in some engineering problems, such as the agitation of bays, estuaries, and harbors. Thus, the waves appear to consist of packets or groups of waves, with smaller waves in the beginning and at the end of each wave packet surrounding larger waves.

Figure 5 shows an example (obtained from a computer simulation derived from a peaked energy spectrum) of the variation of a water surface with time at a fixed location. Note that the wave heights look relatively random, but the wave periods (the time between wave crests) appear reasonably constant. An overhead snapshot at a fixed time would reveal a variation in wave length similar to the variation of the water surface with time at a given location. Surfers' old advice to look for the seventh wave has some truth. This figure illustrates how waves travel in groups. Whether the seventh wave is the largest is certainly open

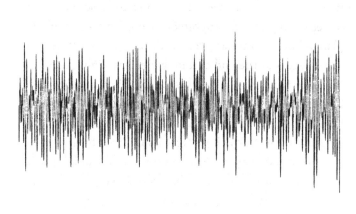

Figure 5 Simulated variation of water surface with time at a given location, showing the "groupiness" of the waves. Time increases from left to right.

to question, but it is evident here that the height of the waves in each group grows to a maximum and then decreases. A musician reading this might notice that there appears to be a "beat" in the wave record.

Is there an easier way to describe these complex groups of waves than by defining the waves by an energy spectrum? This question arose during World War II when it suddenly became very important to describe storm waves simply so that amphibious troop landings could be planned better. Investigators at the Scripps Institute measured waves at the end of a long pier and simultaneously estimated the waves' heights by eye. Remarkably, they found that their visual estimates of waves' heights correlated well with the average of the highest one-third of the measured waves. The investigators referred to this as the *significant wave height*. This descriptor of a train of waves has been used ever since. The coastal wave height mentioned in a weather report is usually the significant wave height; sometimes an estimate of the expected maximum waves will be added. However, the early investigators could have used the average wave height or one of many other "statistical" measures to describe a group of waves.

If waves recorded at a given location were sorted, the record would have a set of wave heights ranging from small to large. To obtain the significant wave height from these sorted data, you would take the average of the largest one-third of the heights. Alternately, a wave height could be

Structures are designed for the extreme—waves recurring at intervals of 50–100 years. Similar to determining the 100-year flood, the coastal engineer constructs a history of major storms that could generate large waves for a particular site.

determined from the height of the waves in the record that are exceeded, say, 1 percent or 10 percent of the time, or whatever percent of time might be chosen.

In designing coastal structures, it is important to define a wave height, usually a significant wave height, that would be expected to recur in a specified period of time—for example, once in 50 years or once in 100 years. Similar to what is done when determining the magnitude of a 100-year flood, a coastal engineer builds a history of major storms generating large waves for a particular site. For example, if the design is to be executed for a wave that is exceeded 10 percent of the time, this wave height is determined for each of the major storms over a large number of past storm years. These data are then used to predict the design wave height corresponding to the year-n storm of interest.

With a lengthy enough sample of wave heights it is possible to construct a probability distribution of these wave heights. A probability distribution of a group of waves expresses the probability that a wave of a particular height will occur in a wave train. In general, the probability distribution of ocean waves doesn't follow a Gaussian (or normal) distribution; rather, it follows a Rayleigh distribution. Whereas the Gaussian probability distribution is bell-shaped and symmetric about the mean value of heights, the Rayleigh distribution is skewed so that the most frequently occurring wave height is less than the mean wave height.

The Rayleigh distribution of wave heights helps to determine the wave height that would be expected to occur with a specified probability relative to the significant wave height. Even though the significant wave height used to define the intensity of waves from a certain storm may seem somewhat artificial, it can be used to determine other wave heights whose proscribed risk of occurrence may be more relevant to a particular problem. In a group of waves, some are smaller than this significant wave and some are larger. From considerations of the Rayleigh distribution the largest wave that one could expect in this random sequence of waves—the probable maximum wave—may be as much as twice the height of the significant wave. Thus, the significant wave is a basis for defining more realistic wave heights for purposes of designing coastal structures. As you wade in the surf and see waves of differing heights coming toward you, although a meteorologist said the waves would be about 3 feet, in reality you may see a wave that peaks at 5 or 6 feet.

Many of our wave data come from instrumented ocean buoys maintained and operated by the National Oceanic and Atmospheric Administration. These buoys were placed near the coasts of the United States so as to be able to report wave data and other environmental information to the public around the clock. The wave data are obtained from measurements of a buoy's motion in waves. (The motion-response characteristics of a buoy are sometimes referred to as its *transfer function*.)

An engineer determining the wave height (or spectrum of heights) to be used in designing a particular coastal structure develops data that will provide the significant wave height (or some other measure of storm wave intensity or spectrum) from a historical set of extreme storms that extend back many years. These data allow the engineer to define a wave to be used for the design of a structure with a predetermined risk—for example, whether it should be designed for waves expected to recur in 50 years or for waves expected to recur in 100 years. The process of constructing data from past storms, usually referred to as *wave hindcasting*, consists of using weather maps from past major storms to predict the characteristics of the generated waves. Sophisticated computer models have been developed to forecast the heights of waves from weather charts such as that shown in figure 20 below. The weather maps used in making predictions may be quite crude. In some cases engineers have gone back to old articles in newspapers published in coastal locations to define historic extreme wave storms. In a conservative design, a probable maximum wave can be obtained from the hindcast significant wave height by assuming that the random wave heights in a particular storm have a Rayleigh probability distribution.

The same methods can be used to forecast waves. For example, given a weather chart for a region in the mid-Pacific, an investigator can estimate the height and the period of waves that could reach a certain location. The time of occurrence of the waves from the observed storm at sea could be estimated on the basis of the distance of the storm from the coast and the waves' predicted speed. Such forecasts warn coastal communities of the potential for dangerous waves from offshore storms.

Since the spectrum of storm waves is usually peaked, with the energy distributed around one frequency (or one wave period), you can conduct the experiment of determining the period of the incoming waves yourself as you sit on a beach looking seaward. A pretty good estimate of the period of the incoming waves can be obtained by dividing the number of breakers you see at a given location into the duration of the observation. It's like taking the pulse of the ocean.

Waves leaving a storm area should be mentioned. In the past, a very simple way to define a storm was to treat the storm region as a "box storm," with the dimensions of the box governed by the shape of the lines of constant pressure (the isobars). The width and length (that is, the fetch) of the "box storm" were prescribed by the region on the weather map where the isobars were reasonably parallel, and the fetch was directed toward the location of interest at the coast. After all, the wind velocity in the region of the storm could be extreme with a very large storm size, but if the wind's direction wasn't toward you the storm might be of little interest to you. (On the other hand, it

Since the spectrum of storm waves is usually peaked, with the energy distributed around one wave period, you can conduct the experiment of determining the period of the incoming waves yourself as you sit on the beach looking seaward. A pretty good estimate of the period of the incoming waves can be obtained by dividing the number of breakers you see at a given location into the duration of the observation. So set your stopwatch and start counting—it's like taking the pulse of the ocean.

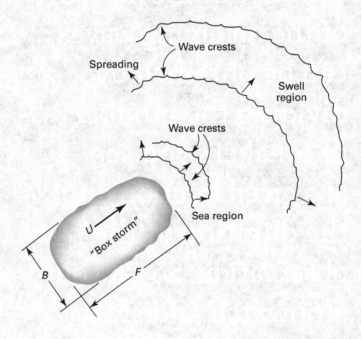

Figure 6 A schematic drawing of the spreading of waves generated by a "box storm." *F* is the storm fetch, *B* is the width of the storm region, and *U* indicates the direction of the wind.

might be of concern to others up or down the coast from you.) The waves generated at the end of the fetch leave the storm area and spread laterally from the end of the "box storm," with wave energy distributed in the wind direction and perpendicular to it. This process is illustrated in figure 6. (You can see the spreading of waves in lateral directions at home by performing a simple experiment: Suppose you move your hand back and forth, generating waves near one end of a bathtub. The waves will be directed toward the other end of the tub, but they will also spread out and strike the sides.) Perhaps even more important, the waves leaving the storm area are complex both in amplitude and in period. In addition, the travel distance can have a considerable effect on the waves that ultimately strike the beach. For example, in Southern California travel distance can result in large differences between the wave height and wave period of summer and winter waves. Sometimes the summer waves from major storms in the South Pacific are more destructive to Southern California's shoreline than the waves generated by strong Alaskan storms in the winter.

When waves are independent of the fetch or the duration, the sea is considered fully developed and the significant wave height and the period of the peak of the spectrum of the waves are primarily dependent on the wind speed. M. K. Ochi (1982) proposed two expressions to approximately describe the height and the period of waves leaving the storm region for this case. In the first expression, the significant wave height is approximately

$$0.21\left(\frac{U^2}{g}\right)$$

—that is, 0.21 times the ratio of the square of the wind speed to the acceleration of gravity.

In the second expression, the wave period corresponding to the peak of the spectrum is approximately

$$7.2\left(\frac{U}{g}\right)$$

—that is, 7.2 times the ratio of the wind speed to the acceleration of gravity. The same units must be used for the velocity and for the acceleration of gravity in both expressions—for example, meters or feet per second for the velocity and meters or feet per second squared for the acceleration of gravity.

The foregoing discussion of how wind generates waves was couched in concepts that are fairly easy to grasp. Wave generation and wave travel from a storm area are certainly more complicated. Computer models of the wind field and the resulting waves developed over the years define wave conditions for engineering solutions to coastal problems for which a simplistic approach isn't warranted.

How Waves Travel and Transform

Wave Travel

We know how waves are generated, but how do they move? This discussion of wave travel will be limited to so-called small-amplitude waves. These waves are not very large relative to the depth (perhaps having a height less than about 10 percent of the depth), and in the deep ocean their height is relatively small relative to their length (the ratio of height to length is less than about 6 percent). With these restrictions, coastal engineers say, these waves can be described by linear mathematics. One of the advantages of using linear mathematics is that waves with different characteristics can be superimposed by simply adding them at a given location. This can be very useful in describing certain events. However, if the ratio of wave height to depth as waves approach the shore exceeds the values stated above, or if in deep water the ratio of height to length exceeds the values stated above, more complex math is required. The waves are then referred to as finite-amplitude or nonlinear waves. The variation in waves' periods in a storm sea usually is considerably less than the variation in their height. For this reason, the shape of an energy spectrum or a wave-height spectrum generally is roughly triangular, with the wave period (or frequency) associated with the most energetic waves corresponding to the peak.

The generation of waves at the "downwind" end of a storm region can, in a crude way, be modeled after a piston extending vertically through the depth and moving back and forth, generating a group or a train of waves. Since there are no walls, the waves will spread laterally as they move away from the generator. Such lateral spreading was sketched in figure 6. As the waves spread and move farther from the source, their height has to decrease. This is because only a certain amount of energy was put into the wave system by the storm. As the carrier of this energy (our wave system) travels away from the generator and spreads laterally, the same amount of energy that was present at the start must be spread over a larger and larger area. Thus, to an extent, the distance of a storm region from a coast determines the intensity of the waves that finally strike the coast.

In addition to the spreading, the waves begin to disperse with distance from the generation region according to their periods. Recall that in deep water a wave's speed is directly proportional to its period, so that with dispersion the longer-period waves move out in front of the shorter-period waves. Thus, whereas in figure 6 the generation region was depicted as a developed sea with a range of wave periods, we begin to see the longer-period waves (or the swell) separate if the distance from the storm to the point of interest is sufficient. Because shorter-period waves tend to be dissipated more readily than longer-period waves, we get the spectral energy distribution for a swell shown in figure 3.

Anyone who thinks that a very distant storm can't cause local problems should consider some waves that sometimes occur along the southern coast of California during the summer. During the winter, large storm-generated waves that arrive in Southern California usually come from storms located in the Gulf of Alaska. On some occasions, extreme winter storm waves in Southern California come from storms near Japan. Those waves propagate eastward, striking the Hawaiian Islands before Southern California. And in Southern California during the summer, longer period waves from large winter storms in the region of New Zealand and Australia occasionally arrive. Although they travel thousands of kilometers before striking the Southern California coastline, they have the potential to cause severe damage to south-facing beaches.

Astronomical tides are dynamic and propagate as water waves. They produce currents at locations that vary with time. They can cause huge time-varying changes in the water surface elevation at particular sites, depending on the shape and depth variations of the coastline. The Bay of Fundy, at the northeast end of the Gulf of Maine, is famous for its very large tidal excursion—about 16 meters (53 feet) from high tide to low tide at the landward end of the bay. Extreme tides also occur at many other locations. One site in the Cook Inlet in Alaska has such a large tidal range that ships docked at high tide can end up resting on the bare bottom at low tide. At the end of one arm of the Cook Inlet, the Turnagain Arm, the tidal range is about 9.2 meters (30 feet); this is the largest tidal range in the United States. The Cook Inlet is one of only about sixty bodies of water worldwide to exhibit a "tidal bore," a condition in which the tide becomes like a breaking wave and propagates up the channel. Under certain conditions during high spring tides, tidal bores can attain a height of more than 1.8 meter (6 feet) and can travel at 24 kilometers (15 miles) per hour.

Tides have periods of 12 or 24 hours, depending on location. (The period of a tide is the time elapsed between high tides.) At an ocean depth of about 4,000 meters (13,120 feet), tides are shallow-water waves. (Remember, waves follow the shallow-water simplification of our equations when their length is more than 20 times the depth.)

Tsunamis also travel as shallow-water waves. The waves that travel from the earthquake region have periods of tens of minutes and lengths on the order of 300 kilometers (200 miles) in the deep ocean. As they propagate in the deep ocean, the waves' amplitudes are very small relative to their lengths. The typical length of tsunami waves is about 100 times the depth. Thus, the speed of travel of the wave in the ocean with an average depth of about 4,000 meters (13,120 feet) would be about 700 kilometers per hour (440 miles per hour)—the speed of a jet plane. Remember that this is the velocity of the "wave shape" (that is, the phase speed), not the velocity of the water particles under the wave. For such long waves the water-particle velocities under the wave are equal to the product of the ratio of the amplitude of the wave crest to the depth and the phase speed:

$$u = \left(\frac{a_c}{h}\right)\sqrt{gh}.$$

For example, with an abnormally large wave amplitude of 61 centimeters (2 feet), a depth of 4,000 meters (13,120 feet), and a wave period of 20 minutes, the water-particle velocity on the bottom would be about 3 centimeters (about 1 inch) per second.

A tsunami spreads from its source and travels great distances before running ashore. In the case of the Sumatra earthquake and tsunami of December 2004, the tsunami caused extreme local damage and loss of life and propagated great distances. Damage occurred as far away as Sri Lanka, the Maldives, and Somalia. As it spread from the source, the length of the crest of the wave increased in a manner similar to what happens to the waves generated by a "box storm" after they leave the storm region or those that spread in a pond after a stone is dropped. Remember that a finite amount of energy is put into the wave system generated by an earthquake, and that is the energy that travels as the wave spreads. However, the spreading of the

The typical length of tsunami waves is about 100 times the depth. Thus, the speed of travel of the wave in the ocean with an average depth of about 4,000 meters (13,120 feet) would be about 700 kilometers per hour (440 miles per hour)—the speed of a jet plane.

wave isn't as simple as that of the waves radiating from the spot where a rock is dropped into a pond. The directivity in its energy propagation and the variation in the depth along its path of propagation both affect the amplitude of the tsunami at a distance from its source. For example, the maximum wave energy of the tsunami generated by the February 2010 Chilean earthquake missed Hawaii because of directivity and because of the variation of the depth along its path. Japan, about 17,000 kilometers (10,700 miles) from Chile, experienced some wave activity from that event.

Because these waves travel across the deep ocean as shallow-water waves, by knowing the depth variation along its path and the fact that the wave's speed is directly proportional to the square root of the depth we have a way of predicting when a tsunami will strike the coast far from the source. This is the basis for our worldwide tsunami warning system.

As waves travel from a deep-water location toward the shore, their characteristics change as the depth changes. However, think for a minute about what happens to the period of a wave as it travels into a changing physical environment. Consider a simple harmonic wave train—that is, a group of waves defined by one wave period. (See figure 1.) Off the Southern California coast the period of a wave may be about 10 seconds; off the Atlantic coast it is less than that.

An example presented by Ippen (1966) is instructive regarding the potential for the period of a simple wave to change as the wave approaches the shoreline. Consider two locations off a hypothetical coastline, the more seaward location denoted as location 1 and the other location as location 2. Allow the depth between these two locations to vary in an arbitrary manner. Now suppose that the period of the waves changes from T_1 to T_2 in traveling from position 1 to position 2. To look at the consequences of this, let's specify the number of waves that enter the region at location 1 as n_1 and the number of waves leaving the region at location 2 as n_2. For this example, in some time interval Δt, the number of waves entering will be $\Delta t/T_1$ and the number of waves leaving the region will be $\Delta t/T_2$. Since Δt is arbitrary, we can make it infinitely large. On the basis of these assumptions, the number of waves accumulated in the region between location 1 and location 2 will be the difference between the number of waves entering the region and the number of waves leaving it—that is, $n_1 - n_2$. Now, if the period of the waves entering the region (T_1) were less than the period of the waves leaving the region (T_2), an infinite number of waves would accumulate in the region. Since that obviously isn't possible, to avoid an accumulation of waves between location 1 and location 2 the period of the waves entering the region must be the same as that of the waves leaving the region. This simple example shows that, although other characteristics of the waves may vary through a region of varying depth, the period of a simple sinusoidal wave train will remain constant as it propagates (travels). If we were to observe a wave at the end of a storm fetch to have a wave period of 10 seconds, the wave period would remain constant all the way to the shore.

What about the other characteristics of waves, such as length and height? Remember that a sinusoidal wave train refers to a group of waves all of the same height, length, and period, as illustrated in figure 1. This simplification allows us to look at a number of changes to the deep-water waves as they travel from the deep ocean to the shore. Considering a wave spectrum, since we can superpose these small-amplitude waves, the same argument of a constant period with distance for each wave component in the spectrum would apply, although the shape of the spectrum might change with time owing to the dispersion of the waves.

Wave Transformation

Wave Shoaling and Breaking

Now let's look at a case in which waves travel toward the shore over a depth that decreases linearly—that is, a case in which the bottom slopes upward, in a direction perpendicular to the coastline, at a constant slope from a given depth offshore to zero at the shoreline. Consider a simple

Waves coming from the Far East cause huge waves at west-facing beaches at popular surfing sites on the north shore of Oahu. These waves can exceed 15 meters (50 feet) in height, even though they have traveled thousands of miles from where they were generated.

wave train, with a constant period T, approaching the sloping bottom from the deep water with its crests parallel to the shoreline. Suppose that at the deepest location these waves are deep-water waves. As the waves travel into shallower water, the period of this harmonic wave group will remain constant. To see what happens as the wave approaches the shore from deep water traveling over this offshore slope, the concept of the energy in the wave is useful. The wave depicted in figure 1 has both potential energy and kinetic energy. Potential energy is the energy necessary to raise or lower the water particles under the wave; the kinetic energy is the energy related to the velocity of the water particles beneath the wave. The total energy per wave length is the sum of the potential energy and the kinetic energy in that one wave length. In an ideal environment, as the wave travels toward shore the total energy in it remains constant even if the wave changes shape. This neglects the effect of bottom friction and the effect of the wind acting on the water surface, both of which may modify the wave's energy. In this simple example, under these assumptions, energy is neither created nor dissipated as the wave travels.

As was shown above, the period of a wave remains constant as the wave travels into shallow water. What about the wave's height and length? Earlier in this chapter, the length of a wave in deep water was defined as a function of the square of the wave's period. What happens to the

Figure 7 A wave breaking at Laguna Beach. (source: author)

Figure 8 A plunging breaker. (source: U.S. Army Corps of Engineers Shore Protection Manual, 1984)

Figure 9 A spilling breaker. (source: U.S. Army Corps of Engineers Shore Protection Manual, 1984)

Figure 10 A surging breaker. (source: U.S. Army Corps of Engineers Shore Protection Manual, 1984)

Figure 11 A collapsing breaker. (source: U.S. Army Corps of Engineers Shore Protection Manual, 1984)

wave's length as it travels into shallower and shallower water? The wave's length is the product of its speed and its period. (Since speed is equal to length divided by time, if we multiply speed by time the result is length.) As the wave travels on the sloping bottom, the wave eventually becomes a shallow-water wave with a speed proportional to the square root of the depth. Thus, the wave's speed decreases as it propagates into shallower water. Since the period remains constant as the depth decreases, the length of the wave (which was initially a deep-water wave) will decrease continuously as the wave approaches the shoreline.

What happens to a wave as it travels in shallow water if its length decreases with the decrease in depth? Since the energy per wave length in a wave (for this simple case of a small-amplitude sinusoidal wave) remains constant even as the wave's length decreases as the wave approaches the shore, the wave's height must increase. This is because we are summing all the energy over one wave length, and in order for the energy in this one wave length to remain constant as the length decreases there must be a corresponding increase in the wave's height. Using this argument for the wave traveling in shallow water, it can be shown that the ratio of the wave's height inshore to its height offshore is equal to the fourth root of the ratio of the offshore depth to the inshore depth.

Another way of understanding the process of the wave's height increasing in shallow water as the wave

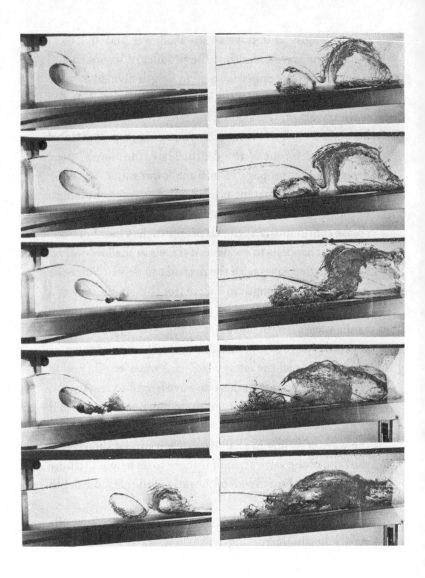

approaches the shoreline is to look at the flow of energy into and out of a simple volume of water near the shore. The volume we will look at is shaped like a box extending in a shoreward direction, with the ends perpendicular to the direction of wave travel. This box is called a control volume. This flow of energy into and out of this control volume, referred to as the energy flux, results in Green's Law, an expression relating the wave's height at the seaward end of the control volume to its height as it leaves the control volume. Green's Law states that in shallow water the ratio of the height of the inshore wave to the height of the offshore wave is equal to the fourth root of the ratio of the water's depth offshore to its depth inshore. Therefore, as the depth decreases in the shoreward direction, the wave's amplitude or its height increases accordingly. (As I have noted, these concepts stretch things a bit, since the waves are no longer what I've called small-amplitude waves as they approach the shore.)

The aforementioned process is referred to as *wave shoaling*. (The wave no longer remains a "small-amplitude" wave in this shallow water, and the amplitude of the crest increases and the amplitude of the trough decreases.)

Figure 12   A laboratory view of a single (solitary) wave breaking on a plane slope, showing the jet forming at the crest, the impact of the jet on the front face of the wave, and the resulting splash-up and run-up of the wave on the slope. (source: author)

Figure 13 A surfer in a large wave tank at the O. H. Hinsdale Wave Research Laboratory of Oregon State University at Corvallis. (source: D. Cox).

When a wave shoals, a maximum wave height is reached, the wave is no longer stable, and something must happen. What happens is that wave breaks and then collapses. Energy is dissipated in the process. We now have a simple explanation of why this wave breaks somewhere near the shore: the wave grows, reaches a limit at which it becomes unstable and cannot "maintain itself," then must collapse. During wave shoaling, the velocity of water particles near the crest of the wave also increases. At some point the velocity of the water particles becomes equal to the wave's speed. Remember that in the simplified approach the ratio of the velocity of the water particles to the wave's speed for a shallow-water wave is equal to the ratio of the wave's amplitude to the depth. Thus, if the wave's amplitude is about equal to the depth, the velocity of the water particles is approximately equal to the wave's speed. If the velocity were to increase any more, the crest of the wave would be traveling at a velocity greater than the wave shape, and the wave would tumble over—that is, break.

Up until now, I have said that a wave propagates in a direction perpendicular to the shoreline. But in reality, more often than not, a wave approaches a beach at an angle. The wave begins to break at a location where the depth has decreased to the point where the wave is unstable. If the wave is traveling at an angle to the depth contour where the wave begins to break (for a plane beach the contours are parallel to the shoreline), the point of breaking will appear to travel along the beach parallel to the shoreline. This is why, as you sit on a beach, you may see white water, delineating wave breaking, travel along the beach parallel to the shoreline.

Wave breaking is a complicated process, as figure 7 shows. Near the center you can see the crest of the wave just beginning to form the familiar jet of a plunging breaking wave. Farther to the right the jet is seen just touching the water shoreward of the wave, whereas to the far left we can see the large "splash-up" that takes place shoreward of the breaking. This splash-up results in large turbulence, and can be another cause of unsafe conditions for swimmers.

Waves do not break only as plunging breakers, as shown in figure 7. Depending on the offshore bottom slope and the wave period and the deep-water wave height, there are different forms of breaking waves—plunging, spilling, surging, and collapsing waves. (See U.S. Army Corps of Engineers Shore Protection Manual, 1984.) Figure 8 shows plunging breakers. A spilling breaker (figure 9) breaks gradually and is characterized by white water at the crest. (Closely observing these waves in the laboratory indicates that these waves may actually begin breaking with a small jet at the wave crest.) Surging breakers (figure 10) build up to break as plungers, but the base of the wave surges up the beach slope before the crest can plunge forward. Collapsing breakers (figure 11) have been defined as a breaking wave in transition from plunging to surging. (Some amateur videos taken in Thailand in December of 2004 show that at some locations a tsunami became a monstrous plunging breaking wave close to shore.)

Breaking can be generated and observed in a laboratory. Figure 12 presents a sequence of photographs showing a single wave propagating up a plane beach and breaking. This solitary wave is like a single hump on the water surface. In the upper few panels, the jet from the plunging wave can be seen striking the front face of the wave. After that there is a large splash-up that gets as large as the height of the incoming wave itself. This is very much like the plunging breaker seen in figure 7, and it was this photo that led to a study that yielded one of the results shown in figure 12.

People about to be inundated by a breaking wave are often told to "dive deep under the wave." This generally reduces the forces on you, as you now are in a region under the wave where the velocities are somewhat reduced. If instead you stand in place and let the breaking wave strike you, the impact force associated with the front face of the wave (now traveling at the wave's speed) can easily knock you down. In fact it can cause serious injury by knocking you off your feet and tumbling you into the splash-up zone.

In figure 7, the portion of the wave on the left has already broken; the portion of the same wave on the right is just in the process of breaking. Thus, to an observer sitting high and dry on the beach, the breaking region appears to sweep up the beach from left to right as a result of the wave's approaching the beach obliquely.

Over the years, much of what has been learned about water waves has been learned by generating, in laboratory wave tanks, waves modeled on real-world observations. In these relatively narrow and long tanks, a movable plate is located at one end. The plates can be moved back and forth in a programmed manner to generate realistic waves that can be used to investigate a wide range of phenomena, including wave breaking. Tanks as large as several hundred meters long with large widths and depths have been used. Investigators sometimes seek out these large facilities so they will not have to reduce the scale of the simulation too much relative to what occurs in nature. A wave tank in the O. H. Hinsdale Wave Research Laboratory of Oregon State University in Corvallis is large enough to minimize some of the effects of scale on laboratory results. That tank—104 meters (342 feet) long, 3.7 meters (12 feet) wide, and 4.6 meters (15 feet) deep—has produced waves large enough to be surfed. (See figure 13.) Even larger wave tanks have been built in Germany and in Japan. Smaller wave tanks are also used in laboratories to investigate a wide range of coastal phenomena. The acceptable tank size (i.e., the length, width, and depth) for a given study depends on the problem of interest and the importance to the study of the effect of scale.

As a wave propagates into shallower water, the amplitude of the crest of the wave increases while the amplitude of the trough decreases. Thus, some studies have looked at a wave just before breaking as simply a hump of water on the surface—that is, a wave with a crest, but no trough, traveling toward shore. These solitary waves are not the simple small-amplitude waves that were discussed earlier; they are finite-amplitude waves, and describing them requires math more complicated than what is required to describe simpler waves. In a solitary wave, the water particles, instead of moving forward and backward in a closed elliptical orbit, move forward and stop. The solitary wave travels at a speed equal to

$$\sqrt{g(h+H)}$$

—that is, the square root of the product of the acceleration of gravity and the sum of the depth and the wave's height. As the solitary wave travels up the plane sloping beach, it grows in height until it breaks. At breaking, the height of the wave is about 78 percent of the depth.

Let's look at a simple example to get some idea of breaking conditions. Suppose a wave breaks in a depth of 1 meter (about 3.3 feet). Using the idea just presented of a solitary wave representing a wave just before breaking, the wave's height at breaking would be 0.78 meter, or 2.6 feet. The wave's speed at that depth would be about 4.2 meters per second (13.7 feet per second, 9.4 miles per hour). The horizontal water-particle velocity would be approximately equal to the wave's speed at breaking—a formidable velocity. On a very steep beach, you might not notice the height of the unbroken wave increasing as the wave approaches the shore, but it may appear to suddenly peak and then break very close to the shoreline. This occurs quite often at Newport Beach in California. At Atlantic City or at Galveston, the offshore slope is very mild and the waves can be seen from afar as they approach the beach, peak, and break (sometimes far offshore).

In a more general sense, both the type of the breaking wave (plunging, spilling, etc.) and the height of the wave when it breaks are defined by the period of the wave, the slope of the beach, and the height of the wave in deep water. If you have a measure of the wave's period generated by a storm far at sea and the height of the wave in deep water off the coast, you can estimate the ratio of the breaking wave's height to the deep-water wave's height. I'll add a bit of complexity here by introducing a parameter: the ratio of the deep-water wave height to the deep-water wave length. Remember, the latter is equal to

$$L_o = \frac{g}{2\pi} = T^2.$$

This parameter is non-dimensional—that is, it doesn't have units. For the transition between surging and

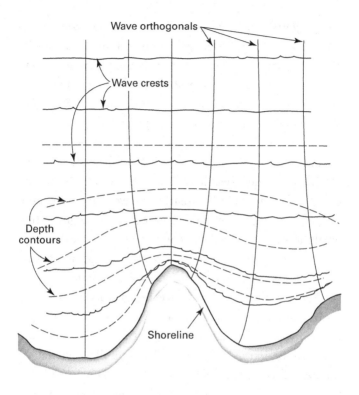

Wave orthogonals

Wave crests

Depth
contours

Shoreline

Figure 14 A schematic drawing of the refraction of a wave as it travels from deep water to the coastline. The bottom contours (the lines of constant depth) are shown as dashed lines. The "wiggly" lines are the crest of the refracted wave at different times as it approaches the shore. Wave orthogonals perpendicular to the wave crests are shown as solid lines.

plunging breaking waves, and for beach slopes with vertical-to-horizontal ratios ranging from 1/10 to 1/50 (a ratio of 1/50 means that the depth increases by 1 meter for every additional 50 meters offshore), the value of the ratio just described ranges from about 0.004 to 0.008. (See 1984 U.S. Army Corps of Engineers Shore Protection Manual.) The height of the breaking wave would be about 1.8 times the deep-water height. If the ratio of the deep-water wave height to the deep-water wave length is between 0.019 and 0.05, we are at the transition between plunging and spilling breaking waves, and the ratio of the height of the breaking wave to the deep-water wave height will be approximately 1.2. (These results are from a large number of experiments conducted in the United States and in Japan.)

Suppose that the slope of the beach is 1 vertical to 50 horizontal (a rather steep beach), that the wave height in deep water off the beach is about 3 meters, and that the wave period is 10 seconds. The non-dimensional ratio mentioned above will be about 0.02, the height of the wave at breaking will be about 1.2 times the deep-water wave height (that is, 3.6 meters), and the wave will be just at the transition between a plunging breaker and a spilling breaker. If you go out into the water up to you knees, you shouldn't be surprised if the breaking wave that strikes you has a height that is roughly as high as the depth of the water you are standing in, and it strikes you at chest height.

Until now I have paid attention primarily to the breaking of waves in shallow water at a beach. However, waves also break in deep water far from the coast. In the deep ocean, wave breaking is no longer controlled by or even related to the depth. It depends on the ratio of the wave's height to its length. At breaking, the ratio of the height to the length will be about 0.14 to 0.17. Suppose the wave period in the deep ocean is about 8 seconds. The wave length will then be about 100 meters (328 feet). Therefore, the wave height at breaking will be about 14 meters (about 46 feet)—about the height of a four-story building. Waves much larger have been observed in the deep ocean. Sometimes referred to as *rogue waves*, these can be caused in several ways. They can be generated by different waves traveling from various directions and meeting at one spot in the ocean, or they can be caused by the influence of major ocean currents on storm waves. Rogue waves are probably more pyramidal in shape than normal wind-generated waves, and they may be higher than 0.14–0.17 times their length.

Ocean liners traveling in the Atlantic and oil exploration platforms in the North Sea have encountered waves as high as 25 meters (80 feet). Off South Africa there has been at least one case of a bulk carrier ship hundreds of feet long breaking in two because of rogue waves. In that case the waves may have been generated by strong currents moving in the opposite direction to the waves.

Figure 15 A view of breaking waves at Turtle Bay, Oahu. (source: author)

Wave Refraction

Waves approaching a plane sloping beach in a direction that is perpendicular to the shoreline have been discussed. But what happens if an incoming wave is moving at an angle to such a beach? Consider a wave with a straight crest line moving in deep water at some angle to the contours of a plane beach. (The contours are lines of constant depth, and for the plane beach I have been discussing these lines of constant depth are simply lines parallel to the shoreline.) As this wave gets into shallower water, its speed is controlled partially by the depth until it becomes a shallow-water wave. At that point the depth alone controls the wave's speed. For simplicity, let's consider only shallow-water waves. As the waves approach the beach obliquely, the part of a wave in deeper water propagating over the plane beach travels faster than the part of the same wave that is traveling in shallower water. The wave crests appear to turn, and as the waves move closer to shore the wave crest tries to become parallel to the contours of the beach. Depending on the initial characteristics of the wave and the beach slope, the crests may not become parallel before breaking sets in. If the wave crest doesn't become parallel to the contours of the beach before breaking occurs, the breaking point of the wave will appear to travel along the contour.

In this simple example, I have essentially described wave refraction—the change in direction of a wave due to the change in its speed caused by the bathymetry (the topography of the bottom). Consider figure 14, an overhead view of what happens to one wave crest as the wave approaches the shoreline. The coastline consists of a headland (a promontory) with a bay to either side. For this example, the bottom contours near the coastline have approximately the same shape as the coastline, and they become smoothed out as one goes offshore. The contours are shown as dashed lines. Far offshore, the wave crests are straight and shown as "wiggly" lines. In addition to curves showing the location of the wave crest as a function of time, the figure shows curves perpendicular to the wave crest at each time. These are called *wave orthogonals* or *wave rays*. For this simple example, let us assume that between adjacent wave orthogonals the wave energy remains constant as the wave propagates shoreward.

Let's look in detail at what happens as the wave approaches the shoreline. First giving attention to the headland, we can see that the wave appears to "wrap around" the promontory, since to either side of the promontory the water is deeper than just offshore of it. Thus, the wave's speed is greater to either side of the headland than just offshore of it, and the wave's crest becomes curved. Since the orthogonals are perpendicular to the wave crest, the wave rays converge on the headland. What happens to the wave height between two rays as these rays converge? Since in deep water two adjacent rays define a certain chunk of wave energy, as we progress shoreward the energy between these two rays will, at least for purposes of this simplified discussion, remain constant. We saw earlier that, because of the change in depth for the case where the rays would be parallel and perpendicular to the straight line depth contours (a plane beach), the wave height increases as a result of shoaling. Now we see that there is another influence on wave height. If the rays converge, the energy gets squeezed between the rays. Thus, the wave height in this region will increase as a result of refraction and the effect of shoaling. Near the headland, the rays converge on the promontory, indicating that the waves become focused there. (This is like the optical phenomenon in which a convex lens focuses light.)

In the bays to either side of the headland in figure 14, we see that the wave rays (the orthogonals) diverge as the wave approaches the shoreline. Thus, considering only refraction, the wave height may decrease in those regions. This is an interesting case, because although refraction decreases the wave height in this region, shoaling of the waves due to the decreasing depth increases it. It is even possible that each effect will cancel the other and the wave height near the shore will be the same as the wave height in deep water far offshore. In the example of figure 14, as the wave approaches the shoreline its crest appears to mimic the bottom contours. Thus, if you want large waves for body surfing, search for headlands, not for coves.

Waves can also be refracted by near-shore currents. When waves travel from still water into a current, as happens at the entrance to a tidal inlet, the waves' speed relative to the still region will change and the waves will be refracted. This is a much-simplified view of a complicated process, since as the waves grow in height the small-amplitude wave approach can no longer be used. A more sophisticated approach is necessary.

Let's look at how variations in the depth offshore and wave refraction can affect wave breaking. Wave breaking can be much more complex than what happens to waves traveling up a plane beach. In most cases the offshore bathymetry isn't simple. The waves may not break in a "line." The breaking may not travel continuously along the beach but instead may appear to occur in "patches." This is probably due both to the character of the incoming waves with discontinuous crests and to the three-dimensionality of the bathymetry. The waves begin to break at a location where the ratio of their height to their depth reaches a critical value, and then the broken wave proceeds shoreward, leaving the foamy trace of breaking behind resulting in "patches." This "patchy" breaking can be seen in figure 15, a photo taken at Turtle Bay on the north shore of Oahu. In the background you can see how the wave has wrapped around a promontory and the breaking wave appears to encircle the point.

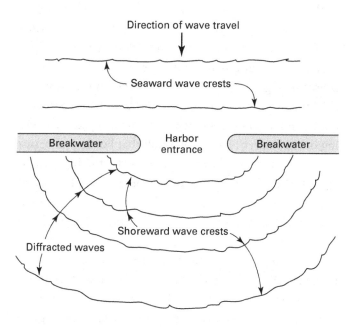

Figure 16 A schematic drawing of a wave diffracting as it passes through the gap in a double-arm breakwater and travels shoreward. The seaward wave crests and the diffracted wave crests shoreward of the breakwaters are shown.

Wave Diffraction

Wave diffraction occurs when waves impinge on an offshore obstruction (such as an island or a breakwater) or on some coastal feature. The obstruction blocks a portion of the waves' energy, and a portion of the energy spreads into the protected region behind the obstruction. In other words, the waves diffract around the obstruction. Since only a portion of the incident wave energy spreads into a protected region, the height of the waves in the region behind the obstruction is reduced. Diffraction also occurs in optics and in sound. For example, suppose you are standing behind a section of a solid wall that borders an active highway. You still hear some sound from the highway, although the sound level is considerably reduced. Some of the sound may be reflected by the wall, but some sound energy is transmitted through and diffracted around and over the wall.

Now consider the effect of diffraction on the waves that propagate past an offshore island and then travel to the mainland. As an example, take Santa Catalina, an island about 35 kilometers (22 miles) from Los Angeles. The island is 35 kilometers long (22 miles) and relatively narrow (13 kilometers or 8 miles at its widest). For this example, consider long-crested storm waves approaching the island in a direction that is generally perpendicular to its longest dimension. Thus, the wave crests are parallel to the long dimension of the island. If there were no diffraction, the island would block the waves completely and the water would be perfectly calm behind the island all the way to the mainland. However, if that occurred there would be a discontinuity in the wave height at either end of the island. To one side of the end there would be waves, but behind the island there would be none. Hence the wave energy flows around the island from both sides to "even out" the heights of the waves at the island's ends. Thus there isn't a calm, wave-free region behind the island; instead, the waves wrap around the ends of the island into the shadow region behind the island. The wave energy that diffracts around the island propagates toward the mainland. Depending upon the offshore wave direction, the heights of the waves along the shoreline may be reduced significantly by wave diffraction around the island relative to their height if the island weren't there. Hence, diffraction provides a degree of protection from extreme storm-generated waves.

The reduction in wave height caused by diffraction depends on the distance shoreward of the island to the location of interest measured in wave lengths. (The distance to the shoreline is measured from the back face of the island, and the wave length is determined approximately by the incident wave period and the depth at the island.) Locating the main town, Avalon, on the side of Santa Catalina facing the mainland has protected it from major storm waves originating seaward of the island. However,

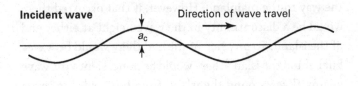

Incident wave Direction of wave travel

a_c

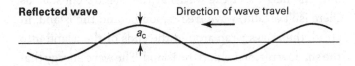

Reflected wave Direction of wave travel

a_c

Standing wave

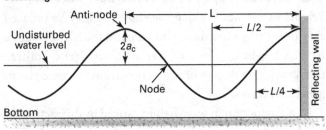

Anti-node

Undisturbed water level

$2a_c$

Node

Bottom

L

L/2

L/4

Reflecting wall

Figure 17 A schematic drawing of a wave incident on a vertical wall, showing the incident wave and the reflected wave at the same instant and the resulting standing wave.

open-sea storm waves diffract around the island, and as a result there is some wave activity at Avalon. And at certain times of the year, strong winds on the mainland known as Santa Ana winds blow from the desert to the coast. During these events, although Avalon is protected from open-sea storm waves by the shadow effect, it is vulnerable to waves generated by Santa Ana winds moving from the mainland toward the island. Small boats and coastal property at Avalon have occasionally been damaged by such wave events.

A somewhat different example of waves diffracting around an island is provided by Flores Island, one of a group of islands extending eastward from Java, Indonesia. Flores Island is roughly conical in shape. An offshore earthquake in 1992 generated a tsunami seaward of the island. When the tsunami struck Flores Island, the very long waves of the tsunami wrapped around the island and met at a fishing village located on the side of the island opposite to where the tsunami first struck. In this case, it was primarily the diffraction of the tsunami around the island that caused significant damage to the village. Similarly, although the town of Avalon on Santa Catalina Island is well protected from normal storm waves, it might be vulnerable to the very long waves of a tsunami. Hence, protection from diffracted waves is related both to the shape of the barrier and to the length of the incident waves.

Breakwaters built to protect or create a harbor are effective because they block a significant part of the energy in the incoming waves and transmit only a portion of the incident wave energy into the harbor. A simple sketch showing waves traveling through the gap between two breakwaters and then propagating toward the shore is presented in figure 16. (Such a breakwater is called a *double-arm breakwater*.) The wave energy that gets through the entrance to the harbor spreads laterally on the shoreward side of the breakwater because of diffraction. As was discussed above, the square of the wave height must be proportional to the energy in the wave per unit length along the wave crest, that is, in a direction perpendicular to the direction of wave travel. Thus, as the total length of the wave crest increases behind the breakwaters, owing to diffraction, the height of the wave per unit length along the wave crest must decrease. In this way, owing to the spread of the wave crest with distance shoreward of the breakwater, the wave height at the coastline is reduced relative to the wave height at the gap in the breakwater. Thus, breakwaters reduce the wave height at the shore in two ways. First, only a fraction of the seaward wave energy is transmitted through the gap. Second, the transmitted energy spreads laterally in the lee of the breakwater because of diffraction.

In the case of a single-arm breakwater or even a headland, diffraction reduces the wave height past the structure in a similar way. In the case of single-arm breakwater, a double-arm breakwater, or a headland, the direction and the period of the incoming waves and the depth define the diffracted wave heights.

Wave Reflection

An obvious function of a breakwater is to limit the wave energy passing a structure. Diffraction (just discussed) is one way this occurs. Another way is the reflection of waves from the structure. We will see later that some wave energy passes through most breakwaters, but here let's consider only the possible total reflection of waves by a structure.

Move one hand forward abruptly to create one wave at one end of a bathtub. This wave (let's call it the *incident wave*) travels down the length of the tub, then reflects from the opposite end and returns to the spot where it was generated. If the opposite end of the tub were vertical, the reflection process would be straightforward. When the initial wave reaches the immovable wall, the velocity of the water particles beneath the surface at the wall has to be zero. If it were not, the wave would have to pass through the wall. To create the condition of zero velocity at the wall, the incident wave generates a reflected wave at the wall such that the velocity of the incoming wave at the wall is exactly canceled by the velocity of the reflected wave at the wall, which now moves in the opposite direction. The incident wave and the reflected wave are illustrated in figure

17 along with the wave that results from adding the incident and the reflected wave that we term a standing wave. In that figure, the incident wave moves to the right and the reflected wave toward the left. Both waves are shown as variations with distance as a "snapshot" in time, each with the same crest amplitude (a_c). Remember that if you stand at one location as a progressive wave like the sinusoidal incident wave sketched in figure 1 propagates you will see one point on the wave (say, the crest) pass you, then one wave period later you will see the same feature pass you. For a standing wave, however, the wave shape does not propagate and varies only with time. At a given location the water surface simply moves up and down with a maximum amplitude that varies from twice the crest amplitude of the incoming wave $(2a_c)$ at the wall and at multiples of half a wave length from the wall to zero at multiples of one-fourth of a wave length from the wall. Thus, there are locations where the water surface doesn't change with time (called *nodes*) and locations where the vertical change is a maximum (called *anti-nodes*). This feature can be seen in figure 17, where the position of the water surface is shown at its maximum. If you have patience, you may be able to see this reflection occurring from a steep beach. There is a place in Maui where, from a comfortable hotel balcony, one can see the incoming waves and then the waves reflected from the beach traveling in an offshore direction.

Owing to other effects, the height of the reflected wave might not be the same as that of the incoming wave. Reflection from a breakwater or from the shoreline isn't as simple as in the tub experiments, since wave energy may be dissipated at the structure; or the wave may run up on the shore and then run down, creating a smaller wave traveling offshore. Wave reflection can cause problems at the shoreline. Consider a seawall constructed to protect coastal property. Waves reflecting from the structure may cause sand to scour in front of the wall, leading to failure of the structure and also to beach damage. Furthermore, the reflection of long waves from a harbor's boundaries can cause significant unwanted currents in the harbor that may delay the loading and unloading of large ships and may damage ships and structures on the shore.

THE WIND

What causes the normal waves we see at a beach? One might just say "The wind." But where does the wind come from?

What causes the wind, and thus what causes ocean waves, is the sun. Energy from the sun, and the rotation of the Earth, give rise to the surface winds that generate waves.

The generation of winds is complicated. In this chapter a simplified and generalized discussion will be presented. Readers looking for more details are referred to Gordon et al. 1998, Neiburger et al. 1982, and Wells 1997.

Some Astronomical Considerations

The Earth's atmosphere extends upward more than 100 kilometers (62 miles). The troposphere—nearly 11 kilometers (7 miles) thick—is its lowest part, extending down

The sun is the energy source that gives rise to the surface winds that, in turn, generate ocean waves.

to the Earth's surface, and is the layer of the atmosphere in which the surface winds that generate ocean waves arise.

The Earth completes an orbit around the sun in 365 Earth days and completes a rotation on its axis in 24 hours. The Earth's north-south axis is tilted at an angle of 66.5° relative to the ecliptic plane (the plane of the orbit around the sun). The orbital plane of the moon is inclined 5.15° relative to the ecliptic plane. (See figure 18.) Owing to the tilt of the Earth's axis, the northern hemisphere experiences summer while the southern hemisphere is experiencing winter. During July, the equatorial region is hot and, although both the North Pole and the South Pole are cold, the North Pole is warmer than the South Pole. Conversely, in December the southern hemisphere experiences summer while the northern hemisphere is experiencing winter and the North Pole is colder than the South Pole. Since the equatorial region remains at approximately the same temperature summer and winter, during the summer, when the North Pole warms, its temperature difference relative the equatorial region is smaller than during the winter.

Generation of Wind

When the Earth heats up in one region, the air in that region expands and thus becomes less dense. As its density decreases, the air in that region rises. Conversely, as the

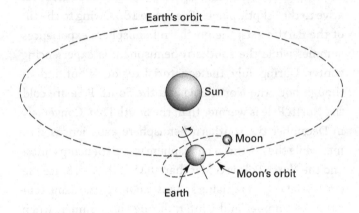

Figure 18 The Earth's orbit around the sun and the moon's orbit around the Earth. The north-south axis of the Earth is inclined 66.5° relative to the orbital plane (ecliptic plane) of the Earth; the moon's orbital plane is inclined 5°9′ relative to the ecliptic plane.

air in a region cools, it becomes denser and sinks relative to the surrounding warmer air. Thus, in simple terms, a *convection cell* (a region in the atmosphere where air moves from one location to another) is created, with air rising over the equator (where the temperature is high) and falling over the North Pole. This convection cell causes pressure gradients at the Earth's surface, with higher pressures in the north and lower pressures in the south. It is this pressure gradient that moves air along the surface of the Earth from north to south.

A simple example of a pressure gradient causing the flow of a fluid is a hose connected to a fire hydrant. At the hydrant, the water is under pressure supplied by the city's water-distribution system. At the open end of the hose (the nozzle), the pressure is that of the surrounding air (atmospheric pressure). The pressure difference between the hydrant and the open end of the nozzle pushes the water through the hose. The larger the difference in the pressure for a given length of hose, the larger the flow. The pressure difference divided by the hose length is the pressure gradient.

The flow of air at the Earth's surface would describe a line on the surface from north to south if the Earth weren't rotating, but of course it is rotating at a speed of about 15° per hour. The rotation gives rise to an imaginary acceleration that was discovered quantitatively in the early nineteenth century by Gustave Gaspard Coriolis. (Its effect had

A simple example of a pressure gradient causing the flow of a fluid is a hose connected to a fire hydrant. At the hydrant the water is under a large pressure supplied by the city's water-distribution system. At the open end of the hose (the nozzle) the pressure is that of the surrounding air (atmospheric pressure). The pressure difference between the hydrant and the open end of the nozzle pushes the water through the hose.

been described qualitatively a century earlier by George Hadley in an effort to explain the trade winds.)

Newton's Second Law of Motion states that force equals mass times acceleration ($F = ma$). This means that if we apply a force to a mass that is free to move it will accelerate in proportion to its mass and the applied force. We can use this expression to define the acceleration in a slightly different way. Newton's Second Law can also define acceleration as force per accelerated mass. The law can be applied to small masses of fluid (air or water) where the rotation of the Earth does not affect the results. The distance over which Newton's Second Law can be applied as if the Earth were not rotating (a *stationary* Earth) is about 100 kilometers (60 miles). However, when the motion of large masses of fluid is considered the assumption of a stationary Earth is a gross simplification. In that case the rotation of the Earth becomes important, and the frame of reference must be from a fixed point in space rather than from a location on Earth. To use Newton's Second Law and a coordinate system fixed on the rotating Earth to treat the motion of large fluid masses, one must add a term that includes the Coriolis acceleration to Newton's Second Law.

To investigate the effect of a rotating Earth, consider the example of a projectile fired from the North Pole directly south toward the equator from position A toward position B' as shown in figure 19. (See Neiburger et al. 1982.) If the distance between A and B' is less than about 100 kilome-

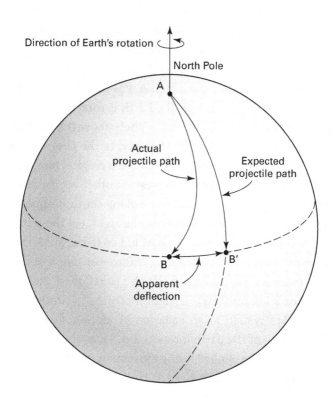

Direction of Earth's rotation

North Pole

A

Actual
projectile path

Expected
projectile path

B

B′

Apparent
deflection

Figure 19   An example of the effect of the Coriolis acceleration on the
path of a projectile fired from the North Pole towards the equator. (source:
Neiburger et al. 1982)

ters (60 miles), the observer at A sees the projectile coming toward position B' on an expected path, tracing a line along a longitude on the Earth's surface. However, if the distance is large (as would be the case for atmospheric movements over a large distance), during the time of travel of the projectile, because of the rotation of the Earth, point B' is no longer directly south of position A. Position B' has moved to position B. Even though the Earth is rotating, the projectile that is moving in the atmosphere above the rotating Earth moves longitudinally. However, to the observer on the Earth's surface at B' it appears as if a magical force has deflected the projectile to the west. In other words, to an observer on the Earth's surface looking south it appears that the projectile is deflected to the right. (In the southern hemisphere, the deflection would be to the left.) This is why the Coriolis acceleration (or force per unit mass) is said to be imaginary: it corrects the equation of motion (Newton's Second Law) written in terms of a stationary coordinate system to an equation of motion that takes the rotation of the Earth into account. Thus, the modified equation can be applied to the movement of large masses of air and/or water over large distances.

Describing the generation of wind over the Earth also entails other complications. One of these—an important one—is the difference between the solar heating of the land and of the ocean. This can cause a confused arrangement of regions of high pressures and low pressures over

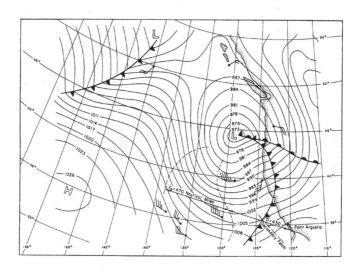

Figure 20   A weather map for 0030 Greenwich Mean Time on October 27, 1950, showing isobars and a suggested fetch for a storm with winds in the direction of Point Arguello, California. (source: U.S. Army Corps of Engineers Shore Protection Manual, 1984)

the oceans. These pressure patterns move around with time, and the storms they create therefore move over the surface of the ocean. The pressure patterns in the atmosphere are usually represented in charts (maps) as lines of constant barometric pressure called *isobars*, with the isobaric pressures usually given in millibars (thousandths of a bar). One millibar is equal to a pressure of 1,000 dynes per square centimeter. In English units, one bar is equivalent to 0.987 atmosphere, or 14.504 pounds per square inch, or the equivalent weight of a 75-centimeter (29.5-inch) column of mercury. Isobars and a suggested fetch for a storm with winds directed toward Point Arguello, California are shown in figure 20—a weather map for 0030 Greenwich Mean Time on October 27, 1950.

How can we determine the speed of the winds that ultimately generate waves from "pressure charts"? The equations that describe the fluid motion are first simplified by neglecting both the nonlinear terms (that is, the convective accelerations) and the term that describes the local acceleration. (In the parlance of fluid dynamics, the equation is linearized and the flow is considered steady.) Under these conditions the force due to the pressure gradient is balanced by the force due to the Coriolis acceleration. Thus, the winds become parallel to the lines of constant pressure (the isobars). The pressure gradient can be obtained from weather chart by dividing the pressure difference between two adjacent isobars by the distance between the isobars.

For these conditions, the wind velocity is directly proportional to the pressure gradient between isobars and is inversely proportional to the product of the density of the air, the rotational speed of the Earth, and the trigonometric sine of the latitude. The velocity determined in this way is referred to as the *geostrophic wind speed*. Imagine that you are standing on an isobar. In the northern hemisphere, if the isobar corresponding to a lower pressure is on your left, the direction of the wind will be away from you because of the Coriolis acceleration. In the southern hemisphere the reverse is the case. When there is a large "pressure gradient" in a region, the pressure-induced velocity is larger than it would be if the same isobars were farther apart. The velocity decreases as the density of the air increases and decreases as the latitude increases (as the north or south pole is approached). This simplified approach breaks down near the equator, as the latitude approaches zero, but it can be used for latitudes greater than about 15° north and south (Gordon et al. 1998). This simplified approach applies to the condition of reasonably straight and parallel isobars.

As an example of determining the magnitude of the geostrophic wind speed, consider isobars with intervals of 4 millibars spaced 300 kilometers apart and air with a density of 1.3 kilograms per meter cubed. At a latitude of 30°, the velocity of the geostrophic wind would be 14.1 meters per second. At a latitude of 60°, the velocity for the same

Before the advent of satellites, wind observations at sea were made using velocity instruments located on a ship's mast, usually taken as 10 meters above the sea surface.

conditions would be 8.1 meters per second. The region corresponding to the geostrophic winds is usually considered to start about a kilometer above the ocean surface. The region below the geostrophic region, is defined as the atmospheric boundary layer. In it, the wind speed must vary from the geostrophic value to the velocity of the boundary (the water surface). (The velocity of the wind and the water must be equal at this interface.) This layer, termed the *Ekman layer*, is where the pressure fluctuations and the shear forces in the wind lead to the generation of ocean waves.

Before the advent of satellites, wind observations at sea were made using velocity instruments located on a ship's mast, usually 10 meters above the sea's surface. The concept of the geostrophic wind obtained from the weather charts of barometric pressures is highly useful in determining the velocities of wave-generating winds. Resio and Vincent (1977) found that the wind speed at mast height is about 40–50 percent of the geostrophic wind speed above about 10 meters per second (33 feet per second). Owing to other effects that will go unmentioned here, at the sea surface the wind direction deviates from the direction of the geostrophic wind by about 10° to 15°.

The approach discussed for defining the wind speed at the water surface allows the engineer to predict wave heights leaving a storm region in a relatively simple manner.

TIDES

The magnitude and the timing of tides are controlled primarily by the tide-generating forces of the sun and the moon. (See Ippen 1966.) But let's look at the mechanism behind the generation of tides more closely. First, we should give some attention to certain important dimensions of these bodies:

The average distance between the centers of Earth and the moon is 384,411 kilometers (238,862 miles).

The average distance between the centers of the Earth and the sun is 149,395,404 kilometers (92,830,000 miles).

The diameter of the Earth is 12,755 kilometers (7,926 miles).

The diameter of the moon is 3,479 kilometers (2,162 miles).

The ratio of the mass of the Earth to the mass of the moon is 81.5.

The ratio of the mass of the Earth to the mass of sun is 0.00000305.

Generally we think of the tides being induced by the moon alone. But since the mass of the sun is nearly 27 million times that of the moon, even though the sun is about 390 times as far from the Earth as the moon is, it is evident that we shouldn't disregard the sun's contribution to the tidal forces acting on our oceans. (The effects of other heavenly bodies on the tides are negligible for purposes of this discussion.)

First let's look at a little geometry. The north-south axis of the Earth is inclined 66.5° relative to the plane of Earth's orbit (the ecliptic plane). The moon is rotating around the Earth in its own plane. That plane is inclined 5.15° relative to the Earth's orbital plane. The moon's orbital plane is inclined at a maximum 28.65° relative to the equatorial plane of the Earth. This is termed the *declination* of the moon, and it comes into play when we consider the variation of the tide with longitude at a given latitude. (The variation with longitude is equivalent to a variation with time due to the Earth's rotation.)

For this part of the discussion only, we will consider the Earth and the moon as producers of the tides. We

Generally we think of the tides as induced by the moon alone. But since the mass of the sun is nearly 27 million times that of the moon, even though the sun is about 390 times as far from the Earth as the moon is, it is evident that we shouldn't disregard the sun's contribution to the tidal forces acting on our oceans.

know from Newton's law of universal gravitation that the attraction of the Earth and the moon to each other is proportional to the product of their masses and inversely proportional to the square of their distance apart. This mutual attractive force must be balanced by the centrifugal force that arises from the Earth-moon combination revolving around a common center. If centrifugal force weren't acting as the Earth and moon revolved, the moon would collide with the Earth. Conversely, if there were no gravitational attraction between the Earth and the moon we would fly off into space. It is the balance of the centrifugal force and the gravitational attractive force that keeps the Earth in equilibrium with the moon. The common axis of revolution of the Earth-moon combination is located between the Earth and the moon. The distance of this point of rotation from the center of the Earth is a function of the ratio of the masses of the Earth and the moon and the distance between their centers. In order for the attractive force and the centrifugal force to be in balance, the point of rotation has to be 4,667 kilometers (2,900 miles) from the center of the Earth. The period of revolution of the Earth and the moon about their common axis is 27.32 days.

Now let's assume that the Earth is covered with a uniform layer of water. Because the moon inclined 28.65° relative to the equatorial plane of the Earth, the moon distorts this sheet of water so that it bulges along the line passing through the centers of the Earth and the moon. Owing to the tidal displacement, the water-covered Earth is a prolate spheroid shape—in essence, a squashed sphere. The undistorted water surface (a constant depth ocean covering the Earth's surface) and the distorted ocean are illustrated in figure 21. The rotation of the Earth beneath this uneven sheet of water gives rise to the oscillation between high and low tides at a given position on the Earth's surface.

A mathematical model has been developed describing the distorted ocean shown in figure 21. (See Ippen 1966.) For a 28.65° angle of declination of the moon, the results show that near the equator (latitude 0°) there are two high tides and two low tides, of equal magnitude, per day. These are referred to as *semidiurnal* tides. At a higher latitude of 60° north, the tide is seen to be *diurnal*, exhibiting one maximum and one minimum (also each about equal in magnitude) per day. At a latitude of 30° the computations showed the tide as semidiurnal with the height of the high and low tides different. (This semidiurnal tidal variation is apparent at Los Angeles located at a latitude of about 34° north.) The variation from diurnal (one high tide per day) to semidiurnal (two high tides per day) is caused by the declination of the moon. If the declination of the moon were zero, the tides would be semidiurnal at any latitude. The actual variation of the tides around the coasts of the world is more complicated than this simple mathematical model shows.

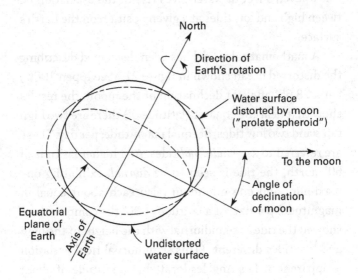

North

Direction of
Earth's rotation

Water surface
distorted by moon
("prolate spheroid")

To the moon

Angle of
declination
of moon

Equatorial
plane of
Earth

Axis of
Earth

Undistorted
water surface

Figure 21 The distorted water surface caused by the Earth-moon interaction and the declination of the moon. (after Ippen 1966)

Even though the sun is much farther from the Earth than the moon is, the huge mass of the sun relative to the moon means that the sun also contributes to our ocean tides. The contribution of the sun to the tides on Earth is 0.457 that of the moon. It takes nearly a month for the moon to rotate about the Earth. In that period, there are two times when the moon and the sun are lined up (the new moon and the full moon) and two times when the tidal effect caused by the moon is in opposition to that caused by the sun (the first quarter and the last quarter). When the moon, the Earth, and the sun are lined up, the tidal force exerted by the moon on the Earth is reinforced by the tidal force exerted by the sun. This results in large "spring tides." When the moon is in its first quarter and when it is in its last quarter, it isn't aligned with the sun, and the tidal effects of the sun and the moon are in opposition. This results in minimum tides for the month, which are called *neap* tides. Since the moon's path around the Earth is elliptical, there are times when the moon is unusually close to the Earth, and at those times extreme spring tides occur.

However, nothing is simple when we talk about coastal effects of ocean systems—either wind waves or tides. The equilibrium theory of tides, although very neat and relatively straightforward, cannot predict with accuracy what the tidal elevation will be at a certain location. Since tides can be viewed as shallow-water waves that travel across

When the moon, the Earth, and the sun are lined up, the tidal force exerted by the moon on the Earth is reinforced by the tidal force exerted by the sun. This results in large "spring tides." When the moon is in its first quarter and when it is in its last quarter, it isn't aligned with the sun, and the tidal effects of the sun and the moon are in opposition. This results in minimum tides for the month that are called *neap* tides.

the ocean, their propagation speed is solely dependent on the ocean's depth. At a point of interest, the tidal height at the coast is affected by the depth variations as the shore is approached, by the dissipation of energy at the bottom caused by friction, by the confining effect of the ocean basins, the influence of the Coriolis acceleration, and by amplification or attenuation caused by local effects of the coastline's shape.

All these effects are evaluated from actual measurements of the variation of the tide with time at a given location taken over an extended period of time. In the United States these local influences are determined by tidal observations over a period of about 20 years (called the *tidal epoch*). Once these data are obtained, tidal prediction becomes possible, since the periods of the various components can be predicted from astronomical observations. For many years these computations were made using a mechanical device (basically a mechanical computer). Now the tidal predictions are accomplished using digital computers and numerical models. These data are assembled so that the magnitude and time of the high and low tides for a given location can be determined a year in advance. Thus, coastal structures can be placed in such a way that they will be unlikely to be harmed by extreme astronomical tidal events.

The Pacific Ocean is deeper and larger in area than the Atlantic. Using a very simple dynamic model, it can be

Since tides can be viewed as shallow-water waves that travel across the ocean, their propagation speed is solely dependent on the ocean depth.

The difference in tide range and in timing on the Atlantic and Pacific coasts of Panama is the reason the Panama Canal has locks to raise or lower ships in transit from one ocean to the other. Only in February are the tides about the same on the Pacific side and the Atlantic side. Although the *average* differences between the Atlantic and Pacific tides at the respective coasts is only about 0.2 meter (0.75 foot), the difference can reach a maximum of 3.7 meters (12 feet) at a given time.

shown that the average mean tidal range (the variation in tidal height) in the Atlantic should be larger than that in the Pacific by about 20 percent. (See Ippen 1966.) On average this is confirmed by measurements. These are very crude calculations, but they indicate that we need more than the simple equilibrium model discussed above to predict the tides with any degree of certainty.

There is large variation in the tidal ranges (the difference between the high tide and the low tide) at locations around the world. In the Bay of Fundy, the tidal range (the difference between the high tide and the low tide) is about 16 meters (53 feet). In the Caribbean Sea, the tidal range is between 10 and 20 centimeters (between 4 and 8 inches).

The difference in tide range and in timing on the Atlantic and Pacific coasts of Panama is the reason the Panama Canal has locks to raise or lower ships in transit from one ocean to the other. Only in February are the tides about the same on the Pacific side and the Atlantic side. Although the *average* differences between the Atlantic and Pacific tides at the respective coasts is only about 0.2 meter (0.75 foot), the difference can reach a maximum of 3.7 meters (12 feet) at a given time.

WAVES AND THE SHORE

Waves on the Beach

Beaches are as dynamic as the ocean that molds them. Figure 22 is a schematic drawing of the profile of a typical beach. This figure shows the nomenclature that I will use to draw attention to various parts of the beach. The formation of the beach and the changes it undergoes are complex. (For more on this subject, see Komar 1998, Dean and Dalrymple 2002, and Bird 1996.)

In this section I will discuss a range of topics dealing with beaches: first a view of what the beach consists, then a consideration of the movement of beach material in a direction perpendicular to the shoreline (called *cross-shore transport*), and then the movement of sand parallel to and offshore of the coast (*longshore transport*). I will discuss various beach protection structures and the concept of beach nourishment.

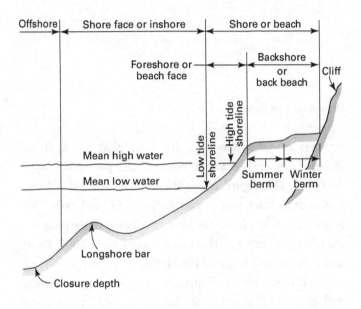

Figure 22 A schematic drawing of the profile of a typical beach, showing the offshore and onshore regions. (source: Wiegel 1964)

The Composition of the Beach

A handful of material picked up from a beach may look very different depending upon where in the world you are. For example, in Nice or in Yalta the material may consist of rounded rock or cobbles centimeters across, in Hawaii it may consist of small bits of coral, and in California it may consist of sand. Beaches in most parts of the world are composed of sand grains from weathered rock located inland, brought to the coast by rivers. The sand at most beaches is beige or yellow, but at some locations, such as along the Panhandle of Florida, the sand is as fine and as white as sugar.

Overseas and in the United States, flood-control dams have been built on many rivers that discharge to the ocean. These dams impede the transport of sediment from the mountains to the coast. This has become very apparent in Southern California, where since 1939 flood-control channels and inland sediment catchment basins have been built to protect life and property. Although these channels and basins have done a good job of protecting the Los Angeles Basin from catastrophic flooding, they have starved the beaches of sand.

The mean diameter of the sand grains on U.S. beaches varies from 0.06 millimeter to 2 millimeters (from 0.0024 inch to 0.08 inch), and the grains vary in shape.

The sand size on the beach affects the porosity of the beach and the ease with which the beach drains as waves

Sand grains are not all the same size or shape, as you and your companion can see from a close examination of a random handful. You would expect a variation in size and shape of the sand grains considering the erosive process that produced the sand in the first place. If you were to build that magnificent sand castle that you were planning, you'd rather not be on a beach with sand that is composed of grains of the same size and shape.

run up and run down its face. As you walk along an apparently dry section of beach near the water, the sand under your foot may take on a wet appearance. As you walk on, that wet footprint may appear to remain wet or may appear to dry up, depending on the beach's porosity. Additionally, the porosity is affected by the way the sand grains pack together. If the grains were all the same size the packing would be much poorer than if there were a range of sand sizes and shapes.

On some beaches, if we start offshore and move toward the shoreline, the sand size increases to a maximum near where waves breaks. The grains are largest at the breaker line; from that location, their size decreases in both the onshore direction and the offshore direction. This is related to the turbulence generated at breaking, which places the sand grains in suspension, and to the onshore and offshore wave velocities that help sort the material. The wave energy acting on the beach has an effect on both the average sand grain size and the slope of the beach.

Suppose that we could find two sand grains, of different sizes but the same shape, that came from the same rock source. If we drop these two grains into the water from the same height, we will see that the larger grain drops faster than the smaller grain. To put it somewhat differently, the larger grain has a higher "fall velocity" than the smaller one. Two particles with the same weight but significantly different shapes also may differ in fall velocity. A breaking wave moves sand from the bottom into the water column, and whether a grain moves onshore or offshore is a function of, among other things, its fall velocity (related to the time it takes the particle to fall to the bottom), the wave period, the wave height, and the depth.

Sand Transport Onshore and Offshore

The water-particle velocity under a non-breaking wave is oscillatory—that is, under the crest of the wave the velocity direction is in the direction of wave travel, and under the trough this is reversed and the velocity direction is offshore. Depending on the magnitude of the fall velocity and the wave characteristics, a sand grain may move either onshore or offshore. (See Dean 1973.) For given wave conditions, larger grains may move onshore, and smaller ones offshore. This is one reason why sand on a beach builds up (accretes) during some parts of the year and "disappears" (erodes) during other parts of the year.

When a wave breaks, the broken wave is projected onshore, bringing sand from the region where the waves are breaking up onto the beach. It is this up-rush and the entrained sand that build the berm shown in figure 22.

What happens to the water that rushed up onto the beach? Some of it may seep into the beach face, but most runs back down the beach slope in the offshore direction. This run-down can travel seaward as an undertow that may be part of a rip current, and if it is strong enough it can carry sand in the offshore direction past the zone where the waves broke. As the water gets deeper, the velocity of the undertow decreases, and a point is reached where sand that has been entrained settles to the bottom. In this way offshore bars can be formed, with the location of the bars essentially a function of the sand size and the size and period of the waves when they broke, see figure 22. Then, depending on later wave conditions, this sand may be moved back toward and up onto the beach. I say "may" because in some cases, such as when waves are generated by an extreme storm, the sand may be moved far enough offshore that normal wind-generated waves can't move it shoreward.

Under some wave conditions, sand with certain characteristics can be moved onshore; under other wave conditions; it can be moved offshore. The sand moved offshore may be "stored" in a submerged bar. In essence, this is like putting the sand in a bank, to be drawn upon at some later time. Under the conditions that moved the sand from the beach to an offshore location, the beach will be degraded (or eroded). When the sand is moved offshore we sometimes refer to the resulting beach profile as a *storm beach* or a *winter beach*.

During the summer, the wave characteristics may be considerably different from those that occurred during the winter, and combined with the sand size the direction of net transport of sediment may be onshore rather than offshore. If the velocities near the bottom and near the offshore bar that was built during the winter are such that material can be transported toward the shore, the "bank" is "opened" and the bar will be a source for rebuilding the beach. The resulting beach sometimes is referred to as a *summer beach* or a *building beach*. Offshore wave characteristics vary from season to season in many locations around the world. In Southern California, winter waves generally are generated by storms in the Gulf of Alaska, and large summer waves arise from storms in the region of New Zealand and Australia. The periods of waves also differ from season to season.

Investigations of the dynamics of beaches have been conducted in the field and in the laboratory. When you look at a coastal problem in the field, you are looking at the actual effect of waves on the coastline. However, sometimes it is difficult to determine why something occurred because of complexities of the coastal wave environment and/or the coastline. Under proper conditions, some of these problems can be studied very effectively in wave tanks and wave basins. As discussed in chapter 1, wave tanks (rectangular troughs) are usually used in two-dimensional studies, with the waves generated not varying across the width of the tank. A wave generator at one end can create waves from deep-water to shallow-water waves, and complex wave trains can be generated that accurately represent the spectrum of waves that are observed in the sea. Wave basins are used in studies in which three-dimensional effects may be important. Wave basins are much wider than wave tanks, and complex three-dimensional bottom and coastal shapes can be modeled in them. Waves can be generated at one end by a series of vertical plates or paddles that can be moved individually so as to generate waves that travel at an angle to the modeled coastline. The waves generated may vary in height and direction across the basin. The response of the waves to the coast or to a coastal structure may also vary across the width of the basin.

When a natural event is studied at a smaller scale in a laboratory, the main question is how the reduced scale of the laboratory model affects the results. Very large wave tanks have been built for the purpose of reducing or eliminating the effect of scale on experiments. The idea is to be able to generate waves that approach the size of the waves seen in the field, but to study their effects under controlled conditions. The best way to investigate a coastal problem is to combine field and laboratory studies.

The location of an offshore bar depends on both the characteristics of the sand and those of the waves. Experiments conducted in the laboratory starting with a plane sloping beach composed of sand showed that the position of the bar offshore was governed by the ratio of wave height to wave length. (See Komar 1998.) Recall that wave length is controlled by wave period for a given location. The experiments showed that increasing the wave height while keeping wave period constant caused the bar to move off-shore into deeper water. In contrast, keeping wave height constant and decreasing wave period (that is, increasing wave steepness) caused the bar to move shoreward into shallower water.

Thus, offshore bars can form in relatively shallow or deep water, depending on the sand size and the wave characteristics. At beaches built by smaller waves, small bars may be prevalent close to shore. These can be very hazardous. In Southern California, a number of swimmers have run into the surf, shallow-dived into the water impacting these bars, and incurred severe injuries. My advice is not to shallow-dive near the shore if you have no idea what the bottom features might be and how they have changed since you were there last.

In figure 22, the berm is the nearly horizontal portion of the beach that is formed by the sand transported by the uprush of waves after breaking near the shore. The broken wave carries sand shoreward with it as it runs up the beach face, depositing it as its velocity decreases. Obviously the highest part of the berm is built by the largest waves in the group of waves breaking in the near-shore region.

Rip currents, strong offshore-directed jets, can be an important feature of some beaches. Various ideas have been proposed about what causes these currents, ranging from the effect of offshore bottom variations to the effects of coastal structures and islands on wave patterns.

When a waves breaks, the broken wave is projected onshore, bringing sand from the region where the waves are breaking up onto the beach.

The velocity of rip currents can range from about 0.5 meter per second to 2 meters per second (1.6 to 6.5 feet per second; 1.8 to 4.4 miles per hour). In addition to picking up unwary swimmers and moving them offshore, these jets carry sand offshore. The jet entrains surrounding water as it travels offshore, and the velocity decreases accordingly. A swimmer who gets carried offshore by a rip current has a better chance of surviving if he doesn't fight the current. Since the velocity decreases with distance from the shore, if the swimmer lets the current carry him offshore he may be able to swim out of it when it becomes weak, then to move parallel to the shore and then back in through the surf to safety.

Sand Transport Along the Shore

In addition to being moved onshore and offshore by waves, sand is also transported parallel to the shoreline. The sediment transport parallel to a beach is called *longshore transport* or *littoral drift*. At some sites this mode of sediment transport can move hundreds of thousands of cubic meters of sand along the shore per year.

When waves break at an angle to the offshore contours, the forces generated have components in the onshore direction and parallel to the shore. These two components form the two perpendicular legs of a triangle; the hypotenuse is the total force. The component that is parallel to the beach contours generates the longshore current. It is

A swimmer who gets carried offshore by rip current has a better chance of surviving if he doesn't fight the current. Since the velocity decreases with distance from the shore, if the swimmer lets the current carry him offshore he may be able to swim out of it when it becomes weak, then back in through the surf to safety.

at a maximum in the region where the waves are breaking (that is, in the surf zone), and it decreases seaward. Tides, ocean currents, and currents caused by the wind generally usually aren't as important in this process. Sediment entrained in the water column during wave breaking will be carried by this current along the beach as well as onshore.

If the waves change direction with the season, the direction of the literal drift also changes over the year, since the longshore transport is a function of the angle of wave breaking to the depth contours in the surf zone. Thus, at any beach where there is longshore transport, there is a net transport rate per year and direction. Table 1 gives the rates of littoral drift for a few selected locations.

As is evident from table 1, sediment transport rates and directions of net transport vary widely around the world. This would be expected in view of the worldwide variation in wave heights, wave direction, and sediment characteristics. Note that all the transport rates in table 1 are for years preceding 1949. Since then, at some sites, there has been a drastic change in the availability of sediment delivered to the beach. In Southern California, inland flood-control projects have significantly limited the supply of sand, which in the past was carried from the mountains to the ocean. Of course waves haven't stopped, and they still have a propensity to move sand. Seasonal movement of sand from one portion of the coast to another can cause erosion in one location or accretion in another.

Table 1 Longshore transport rates at selected sites. (source: Komar 1998)

Location	Transport rate (m^3/year)	Predominant direction	Years of record
Ocean City, N.J.	306,000	South	1935–1946
Galveston	334,700	East	1919–1934
Santa Monica	207,000	South	1936–1940
Port Said	696,000	East	
Durban	293,000	North	1897–1904
Madras	566,000	North	1886–1949

The effects of longshore sand transport have been observed over the years at Santa Monica, where an offshore breakwater parallel to the beach was completed in 1934 to provide a safe haven for fishing boats and to protect an amusement pier extending from the coast. The breakwater was damaged by a storm in the late 1930s so that at high tide the crest of the breakwater was about at the water surface, and it was never repaired. Owing to wave diffraction, an offshore barrier such as this breakwater provides a "shadow" region, with reduced wave energy in its lee. On the average sand, is transported southward in Santa Monica Bay. When sand gets to the region behind the breakwater, there isn't enough wave energy available to move all of it farther south, and so the shoreline grows seaward. The accreting beach takes the form of a tombolo (a "bell-shaped" curve). If there were no wave energy behind the breakwater, the tombolo would grow seaward from the shore and would connect the shore to the breakwater, completely blocking littoral transport through the region. This has happened behind a small offshore breakwater at Venice, a few miles down the coast. However, at Santa Monica the breakwater is porous, and wave energy leaks through over and around it, limiting how far the tombolo extends from the shore.

In March of 1983 a major storm destroyed a portion of the Santa Monica breakwater and about one-third of the amusement pier. As a result of the damage to the breakwater, the wave energy in its shadow was greater than it had been before the storm, and the tombolo that was relatively stable since the 1930s receded significantly. This led to an increase in down-coast littoral drift, and beaches south of Santa Monica that were, to some extent, starved by the effects of the Santa Monica breakwater now receive some of the sand that had been "stored" up the coast for many years.

Controlling Beach Erosion
In many cases, coastal structures built to protect or create a harbor have seriously affected nearby beaches. The erosion of a beach not only affects recreational uses; it

Figure 23   Erosion at Dockweiler Beach, January 19, 1988. (source: author)

Figure 24   Accretion at Dockweiler Beach, California, May 19, 1988. (source: author)

also affects coastal property. At some locations, the beach and the shoreward dune fields protect buildings located near the coastline. Both major storms and normal seasonal waves have caused havoc for coastal communities. At Santa Barbara, a breakwater constructed in the 1930s created a small harbor for boats. At that location, the direction of sand transport is toward the south. As soon as the breakwater was finished, sand began to collect on the upcoast portion of the breakwater. In addition to the sand accumulating on beaches north of the breakwater, sand was transported southward along the seaward face of the breakwater, carried around the tip of the breakwater, and deposited in Santa Barbara Harbor. As a result, the amount of sand transported south along the coast was reduced, and beaches south of Santa Barbara were starved of sand. Erosion threatened buildings near the shore. Accretion and erosion caused by coastal structures are observed at many places where a jetty constructed to protect the entrance to a harbor, a bay, or an inlet has interrupted the longshore transport of sediment. At Santa Barbara the problem was solved by stationing a dredge near the harbor's entrance and pumping the sand collected near the terminus of the breakwater down the coast to the sand-starved beaches.

Channel Island Harbor and Port Hueneme, down the coast from Santa Barbara, are about 1.6 kilometer (1 mile) apart. Before the construction of Channel Island Harbor, the deep canyon off Port Hueneme was a trap for

Accretion and erosion caused by coastal structures are observed at many places where a jetty constructed to protect the entrance to a harbor, a bay, or an inlet has interrupted the longshore transport of sediment.

southerly-moving littoral sediment. It starved the down-coast beaches of sediment transported from the north, creating serious coastal erosion. To alleviate this problem, an offshore breakwater was constructed parallel to the coast at Channel Island Harbor located just to the north of Port Hueneme. It is 700 meters (2,300 feet) long and 600 meters (2,000 feet) offshore in a water depth of 9 meters (30 feet). It creates a "sand trap" behind it similar to the one at Santa Monica. Periodically this sediment is dredged and deposited on beaches south of Port Hueneme to compensate for some of the coastal erosion that had been experienced.

As Komar (1998) pointed out, there are four ways to attempt to alleviate coastal erosion: to do nothing, to retreat and relocate, to build structures to stabilize the coast, and to use beach nourishment. Komar refers to the last two as the "hard" solution and the "soft" solution, respectively.

Doing nothing is tempting, but usually isn't acceptable. In general, at the shoreline, the engineer is faced with a need to mitigate erosion so as to protect the beach or structures located nearby. Taking no action doesn't alleviate the problem, but it may be the appropriate approach when there is no hazard in "letting nature take its course" and when the expense of attempting to reduce beach erosion isn't justified.

Retreating may be the best solution in certain settings. In a number of cases, relocating a coastal structure was the most sensible way to deal with erosion of the shoreline.

This approach can be very expensive, but where it important to save a structure (say, for historical reasons) it may
be the best solution. For example, the historic Cape Hat-
teras Lighthouse was in the path of an eroding beach on
the Outer Banks of North Carolina and was in danger of
being undermined by waves from extreme storms. A deci-
sion was made to save it by moving it. The solution was
ingenious but costly. In 1999 and 2000, the lighthouse,
61 meters (200 feet) tall, was moved, on rails, 870 meters
(2,870 feet) inland.

Now let's consider a few "hard" solutions to beach erosion. (I have already mentioned two instances of this approach: Santa Barbara and Port Hueneme.)

Seawalls and revetments are the most frequently used
"hard" solutions. The object is to construct a wall that will
deflect extreme waves. The wall may be vertical or sloped.
Among the materials that may be used are timber and concrete. A concrete wall generally is sloped in a shoreward
direction, with a curved section near the top. The curved
section throws the wave that runs up on the face of the
wall back toward the sea. Figure 25 shows a laboratory
model of a sloping seawall with a curved upper section.

A seawall may be quite effective in protecting coastal
structures located behind it. However, the combination
the incident waves and waves reflected from the wall may
cause erosion of the beach fronting the structure. To reduce wave reflection and stabilize the seawall while miti-

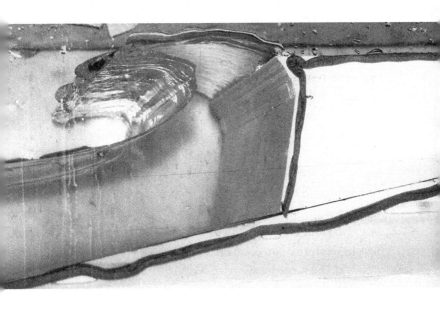

Figure 25 A photograph, taken in a laboratory, showing the reflection of a wave from a seawall with a curved upper portion and the resulting seaward directed jet. (source: author)

gating erosion, revetments are sometimes constructed in front of such walls. A revetment is simply a sloping structure composed of large rocks or various shaped concrete units. In addition to this use, revetments have been built along beaches to protect shoreward located structures.

Among the other shoreline structures that have been built to mitigate beach erosion, sometimes successfully and sometimes disastrously, are groins. These are structures composed of rock, timber, or sheet pilings (or a combination of these) built perpendicular to the shoreline and extending seaward, perhaps as far as 200 meters (650 feet). They are built in pairs with spacing about four times their length (see Komar 1998). They don't affect the incoming waves. The objective is to trap a section of the beach between groins so that sand only moves back and forth between them. Figures 23 and 24 show the recession of the beach in front of a sheet-pile and rock groin at Dockweiler Beach, California, on January 19 and the accretion of the beach at the same location on May 19, 1988. Under optimum conditions, the longshore sediment transport moves past the tips of the groins so that the beaches down the coast aren't starved. However, on a number of occasions, although the beach region protected by groins is reason-

Figure 26   Tsunami measurements at tide gauge stations along California, Oregon, Hawaii, and Johnston Island during the 1992 Cape Mendocino earthquake. The time and amplitude scales are the same for all locations. (source: McCarthy et al. 1993)

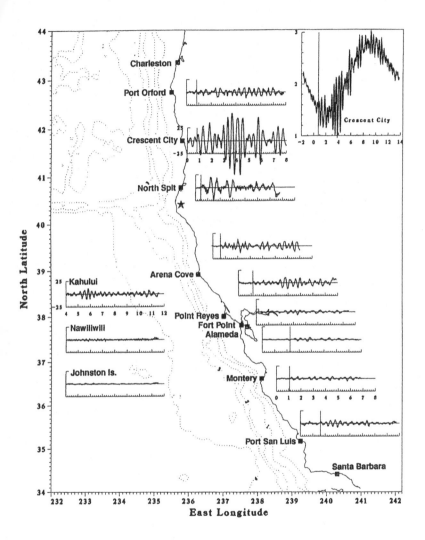

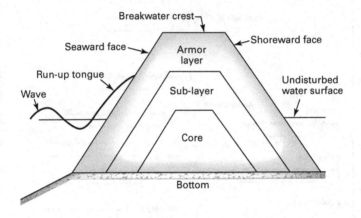

Figure 27 A schematic drawing of the profile of an idealized breakwater.

ably stabilized, beaches down the coast have been starved and eroded.

Relatively short offshore breakwaters built parallel to the coastline have been constructed at some locations to stabilize the beach. The breakwater forms a "quiet" zone behind it where the movement of sand along the coast is interrupted and the sand is deposited behind the breakwater. If the breakwaters are built close enough to the beach, the beach can build out seaward to connect with the breakwater blocking the down-coast transport of sediment. At a number of locations on Spain's Mediterranean coast, offshore breakwaters and groins have stabilized beaches and enhanced their recreational use.

A "soft" solution to beach restoration and/or protection is beach nourishment, in which beach material (either from onshore, from near the shore, or from offshore) is transported mechanically onto the beach to repair or build it. After nourishment, the beach will develop an equilibrium profile (see Dean 1991) such that its cross-section shape develops as a function of sand size and wave characteristics. This method of beach protection has been successful in Miami, Florida, in Ocean City, Maryland, and in many other places. Since this method generally adds sand to an eroded or eroding beach, the results may only be marginal successful and generally need replenishment after extreme storm activity. Thus, it may not be an absolute cure for an erosion problem, but if the cost-to-benefit ratio is reasonable it can protect a beach and coastal structures for a number of years.

Generally the sand used in beach nourishment should be similar to that of the existing beach or even coarser. If the sand is too fine for the location and it is suspended by breaking waves in the surf zone, it may be carried offshore by the waves. Aside from sandy beaches, beaches composed of cobbles and gravel-size material also are candidates for nourishment. Depending on the situation, any of several methods of placing the material may be appropriate (Komar 1998). In dune nourishment, sand is placed shoreward of the beach in the backshore region. This nourishment results in direct protection of shoreward structures. Another option is to place sand on the visible beach (that is, the portion of the beach above the mean water level). A third approach is to place sand over the entire beach profile—onshore and offshore—out to the *closure depth* (the offshore depth beyond which sand is not transported cross-shore), seaward of which the beach profile would be considered historically stable. The last method is to place the material in the shallow offshore waters as an artificial bar and allow wave activity to move it in an onshore direction.

Houston (1996) discusses the importance of beaches to the US economy, and the concomitant importance of maintaining the beaches. He points out that "U.S. beaches are a leading tourist destination" for both US residents and international travelers, and that "85% of all U.S. tourist revenues are earned by coastal states largely due to the attraction of beaches." These figures offer strong support for attention to intelligent beach management.

Long Waves in Bays and Harbors and Their Effect on Shipping

Wave-Induced Ship Motions

If you can gain entrance to a major port near where you live, try to observe the loading and unloading of a large container ship. The containers are metal boxes, some 6 meters (20 feet) long and others twice that size that hold an assortment of goods from ports worldwide. Ships are loaded and unloaded using immense gantry cranes that travel along the wharf on rails that are parallel to the edge of the wharf. The cranes lift the containers from trucks on the wharf, move them in a direction that is perpendicular to the rails on the wharf, and stack them in the ship's hold and on its deck. At each of the four corners of each container are devices that lock the container to the one below it. This clever arrangement allows containers to be stacked five or six high. However, the containers have to be positioned carefully within inches.

Now suppose that waves cause the moored ship to move. Although the crane can move relatively quickly vertically and in the direction perpendicular to the edge of the dock, its size makes it relatively slow in moving along the dock parallel to the long axis of the ship. If because of these motions the ship can't be loaded and unloaded rapidly at one port, its owner may opt for another port. In addition, if a ship moves too much, there is the risk of injury to personnel and damage to the ship and/or the dock. Thus, a port presented with this problem will strive to fix it rapidly. This problem is exacerbated if the harbor in question responds resonantly to waves with the same period as the resonant periods of the moored ship.

When moored to a dock, a ship 300 meters (nearly 1,000 feet) long, with a displaced weight of about 75,500 metric tons (83,200 tons), can be moved by the action of waves that might be only a fraction of a meter high, and its mooring lines can be snapped by the motions that these waves cause.

A moored ship and its mooring lines are, in many ways, similar to a simple spring-mass system in which a mass is connected to a spring that in turn is fixed to a stationary support. If the mass is deflected downward and then released, it oscillates up and down at a particular frequency, termed its *natural frequency*. This frequency is determined by the strength of the spring and the magnitude of the mass. If the mass were to be grabbed and moved at its natural frequency, the system becomes resonant and the motion of the mass would be excessive.

When moored to a dock, a ship 300 meters (nearly 1,000 feet) long, with a displaced weight of about 75,500 metric tons (83,200 tons), can be moved by the action of waves that might be only a fraction of a meter high, and its mooring lines can be snapped by the motions that these waves cause.

We have to be careful when comparing the elasticity of mooring lines to a simple linear spring. For a simple linear spring, the ratio of the applied force to its deflection is constant and this ratio is independent of the displacement. Generally the mooring lines' response to an applied force is nonlinear—that is, the ratio of the applied force to the deflection varies depending on the applied force. Thus, the response of a ship to an excitation is more complicated than the simple model discussed here, but this model describes the important features of wave-induced ship motions.

If you could move a ship a given distance in the fore or the aft direction (the surge direction or the longitudinal direction, in ship parlance) and then release it, the ship would oscillate back and forth at its natural period in a fashion analogous to that of the mass-spring system just discussed. This is a simplified view of the problem, since the elasticity of the mooring lines is nonlinear. The frequency with which the ship moves (its natural period) is controlled by the mass of the ship and the elasticity of the mooring lines. The stiffer the lines are, the higher the frequency (or the smaller the period) of the ship's motion. For ships of the size mentioned above, one might expect periods of motion back and forth along the dock of about one to two minutes. Such ships also may move onto and off the dock (sway) and up and down (heave), in addition to their angular motions of pitch, yaw, and roll. These latter motions are excited by waves with smaller periods, on the order of tens of seconds, arising from more normal offshore waves. Apparently the cranes can handle these smaller-period (higher-frequency) motions, but they have difficulty moving along the dock to keep up with the longer-period surge motions while still being able to accurately place containers atop one other.

During a storm at sea, the periods of the waves that are obvious to someone sitting at the shore are measured in seconds. Waves from extreme storms entering a harbor may cause damage to a port, but they generally do not cause significant fore-and-aft motions of large ships or, at the extreme, resonance of the ship-mooring system and the harbor. Since the most serious effect on loading and unloading occurs if the wave periods are on the order of minutes instead of seconds, the question that arises here is "What is the source of these very-long-period waves?" The answer is complicated and is connected to what happens to the "random" waves with relatively short wave periods on the order of seconds (that is, the irregular wave train) as they approach the shore. Recall figure 5, which showed an irregular group of waves in deep water. The *groups* of waves have periods that are much larger than those of the individual waves in the groups. It has been proposed that in some harbors these wave groups and their nonlinear interactions are what produces long waves (shallow-water waves) with periods of one to two minutes. However, there are other possible generation mechanisms for these long waves. For example, a traveling atmospheric pressure disturbance associated with a storm may cause coastal waters offshore of a harbor's entrance to oscillate and create damaging long-period waves inside a harbor.

If there is wave activity in a harbor that is long period but with small amplitudes, how does it move a ship? It is worthwhile to digress a bit to discuss this, since understanding the forces imposed on structures by waves is a topic that is also important in the design of coastal and offshore structures.

There are two primary forces at work underwater acting on and moving a moored ship. The first is the *drag force*, which is proportional to the square of the water particle velocity and which depends on the shape of the ship's hull, its roughness and the fluid properties. Generally, the more important contribution to the total drag of a ship is the *form drag*, which is a function of the shape of the ship. For example, an ocean-going container ship would probably have significantly less form drag than a rectangular barge with the same displaced weight. The friction drag on the hull is generally less that the form drag. In addition to the underwater forces imposed on the ship by waves, the containers stacked on the deck of the ship become a large bluff area to the wind. The drag due to the wind acting on the stacks of containers on the ship's deck, just like the wind acting on a sail, can contribute significantly to the forces in the lines that moor the ship to the dock. However, these forces generally aren't oscillatory and only affect a ship's dynamics in their effect on the restraining force of the mooring lines due to their nonlinear elasticity.

Form drag can be reduced by streamlining the hull. At significant costs, ships are modeled and tested in laboratories by towing them in large water-filled channels, often with waves, in an attempt to minimize their form drag. Computer-aided design is used today as an adjunct to the testing of a physical model.

At a given location, the velocities under a wave change with time. Thus, there is an acceleration (and deceleration) of the water particles under the waves to which a moored ship is exposed. This accelerating (or decelerating) flow leads to a second force acting on the moored ship that we refer to as the *inertia force*. The inertia force is caused by the ship's modifying the kinetic energy of the fluid. The inertia force is a function of the shape of the ship. Therefore, if we are interested in the variation of the total force acting on the ship, and thus in the forces on the mooring lines and the motion of the ship at its berth, we must consider both the drag force and the inertia force. The variation of these forces with time is important with regard to the potential for resonance of the moored ship. If the period of these time-varying forces is near that of one of the natural periods of the surge motion of the moored ship, resonance can occur and large ship motions can result.

Can resonant ship motions be minimized? Perhaps one way is to change the mooring system so that the natural frequency of the ship is changed. In our spring-mass example, this amounts to stiffening or softening the spring system. But that entails having an engineer specify to a ship's captain how he must moor his ship at a particular location in the harbor—something that a captain isn't likely to accept. A change in mooring location may, in some cases, mitigate the problem. In the extreme, the activity of long waves in the harbor has to be modified so as to reduce the "forcing function" and thus to reduce the ship motion. The latter possibility may be easier to realize when constructing a new harbor than when correcting an existing one.

Wave-Induced Harbor Motions

Related to the activity of long waves in a harbor, let's consider another part of the problem of wave-induced ship motions that may arise in some harbors: resonance of the harbor caused by long-period waves where the periods of the waves are close to the natural period of the moored ship.

A simple example of this type of resonance is the sloshing motion of coffee in a cup moving back and forth. The period of the sloshing motion would probably be the period of the lowest mode of oscillation of the coffee-filled cup. In the sloshing mode, a diameter is the nodal line, and the antinodes are at the surfaces of the cup opposite the node. In a sense we can look at the lowest mode of oscillation of a circular harbor as analogous to the sloshing of the coffee. Instead of the excitation being the motion of the harbor, now the harbor is excited by exposing the entrance to waves of just the right period to create this resonant condition. A simplified explanation of resonance in this harbor is that a portion of the incident wave energy becomes trapped in the harbor with each incoming wave of the "right" length.

Another analogy to harbor resonance may be easier to understand. Consider the motion of air in one of an organ's pipes. The pipes are of different length so that each will resonate at a different frequency. Changing the length of a pipe would change the frequency of the sound. In a similar way, changing the length or the shape of a harbor would change the frequency or period that causes the harbor to resonate. It is, of course, difficult and expensive to do this with an existing harbor, but under dire circumstances it may be the only way to reduce excessive ship motions.

Modifying the shape of a harbor so as to minimize the effect of the harbor's resonance can reduce the velocities and the accelerations to which a moored ship is exposed. This was done (at significant expense) in the 1990s at a berth in the California port of Long Beach. (See Poon et al. 1998.) The problem of ships' moving while at the berth had become so severe that at times it was necessary to have tugboats press a ship up against the fender system of the dock to prevent motion and to permit loading and unloading. Without help from tugs, the surge motion of a ship moored at this berth caused the steel mooring lines to cut through the hull where they exited the ship, creating a hazard to personnel on the ship and on the dock.

The particular berth where this occurred is nearly rectangular, with generally vertical sides. Very long waves with periods on the order of minutes created resonance in the harbor, with wave periods that were close to the resonant periods of the moored container ships. Extensive experimental and theoretical studies showed that a system of breakwaters creating an outer region to this berth reduced the wave-induced oscillations in the harbor. The breakwaters reduced the incident wave energy entering the berth. In addition they changed the shape of the harbor thus changing its resonant periods. This was a $21 million solution. The breakwaters created a more tranquil basin, and ships have been unloaded safely ever since they were built. Another factor helping the situation has been an increase in the size of the ships using this berth. Larger ships result in a mismatch between the ship-surge periods and the resonant periods of the basin.

Two other examples of harbor resonance. are provided by the Bay of Fundy (on the northeast end of the Gulf of Maine, between the Canadian provinces of New Brunswick and Nova Scotia) and Crescent City Harbor (in Northern California).

Resonance in the Bay of Fundy is a natural phenomenon that occurs because the period of the tide is close to a resonant period of the bay. Near the landward end of the bay, resonance results in a tidal range of 16 meters (53 feet). The oscillation has been proposed by some engineers as a possible source of energy from the sea. If the increase in the water level at the coast caused by the excessive tides could be captured in a reservoir, the stored water could be used to generate electric power at low tide.

In Norway, the occurrence of resonance in a man-made coastal structure was proposed as another way to generate electricity from wave energy. The idea was to build a circular basin at the coast, with a gap opened to the sea, that could be resonantly excited by normal wave activity. The circular basin was to be capped. The oscillating water surface in the basin would force the air above it through a turbine connected to an electrical generator. This has limitations due to the generally changing wave energy spectrum from day to day.

Crescent City is periodically attacked by both local and distantly generated tsunamis. The lowest resonant period of Crescent City Harbor is about 22 minutes (Lee et al. 2008). This is too large for excitation by normal sea waves or by the "groupiness" of these waves, and too small to be excited by the tides. However, it appears to be about the right period to be excited by tsunamis. It takes a certain amount of time, measured from the impact of the first wave, to put a bay or a harbor into resonance. Thus, the maximum excitation of a bay or a harbor isn't reached instantaneously. At Crescent City, some of the largest waves caused by tsunamis occurs hours after the first wave reaches the harbor. When the tsunami generated in the Kuril Islands on November 15, 2006 hit Crescent City, the most destructive wave occurred two hours after the first wave reached the harbor. This delay in large wave activity creates danger for people who may go to the harbor after the first waves are observed to see how much damage has occurred.

An excellent example of the resonance induced in harbors by a tsunami is provided by the 1992 Cape Mendocino earthquake. McCarthy et al. (1993) propose that the earthquake resulted from slippage on the southern end of the Cascadia Subduction Zone. Figure 26 shows tide-gauge records (with the tide removed) from locations in California, Oregon, Hawaii, and Johnston Island on April 25. (The earthquake's location is indicated by the star.) The record of the tsunami at Crescent City, shown in the inset at upper right, reveals that the tsunami arrived at low tide. Because all the tide-gauge records are plotted with the same time and amplitude scale, they can be compared directly. At each location, the time of the arrival of the tsunami is indicated by a vertical line in the record. The most obvious feature of the figure is the significantly greater amplitude of the tsunami at Crescent City relative to the amplitude at the other harbors. This appears to be attributable to the resonance of the harbor at Crescent City, described by Lee et al. (2008).

Never walk out on a coastal structure during a storm. You never know when a large wave might sweep over the structure.

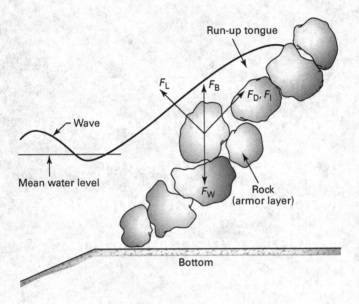

Figure 28 A schematic drawing of a rock resting on a nest of rocks on the seaward face of a breakwater, showing the forces to which the rock is exposed during wave run-up. F_D is the drag force, F_I is the inertia force, F_L is the lift force, F_W is the weight in air, and F_B is the buoyant force.

Breakwaters

At first glance, a breakwater, a jetty, or a groin extending from the shore or located offshore may appear to be merely a pile of rocks. Indeed, some of the older breakwaters were just that. However, modern breakwaters, jetties, and groins are carefully engineered structures.

Some General Considerations

Breakwaters are built to protect harbors, bays, and the coastline by reducing wave activity, and sometimes to create a harbor along an open coast. The use of rock structures to protect harbors goes back thousands of years. One of the earliest known harbor-protection schemes was devised around 2000 BC for the Port of Pharos on the open coast of Egypt. It entailed a breakwater built of rock approximately 2.59 kilometers (1.6 miles) long, with smaller stones placed in the spaces between large blocks of stone to make the breakwater more stable (Savile 1940).

Revetments built at the coast of rock or concrete units protect structures and the coast itself. When placed correctly these can prevent seaside cliffs from collapsing. Jetties (breakwater-like structures constructed seaward from the entrance of a harbor) generally protect a harbor's entrance from waves coming from directions that are at an angle to the entrance. Since they extend seaward, they can "catch" sand being transported parallel to the coast and prevent it from being deposited at the harbor's entrance. Of course, such structures may cause significant beach erosion down the coast. They are constructed much like breakwaters. And groins, as has already been noted, are short structures, usually extending perpendicular from the beach, built in groups to reduce overall beach erosion. Revetments, jetties, and groins are usually constructed of rock or of a combination of rock and other materials. Even though these various kinds of structures are built for different reasons, all of them are constructed in similar fashion.

An ideally designed breakwater forms a protective wall that blocks waves from entering a harbor except through an opening of limited width. Offshore waves are reflected from the breakwater and diffract through the harbor's entrance; as a result, the heights of waves inside the harbor are greatly reduced.

The two breakwaters built in the late 1990s to change the resonant characteristics of a berth for container ships in Long Beach Harbor had a total length of about 1,000 meters (3,300 feet). The cost of building them was about $21,200 per meter.

An even longer breakwater was built over a period of decades to protect the adjacent ports of Los Angeles and Long Beach. Its three sections form an outer harbor for the two ports with entrances to each port. The total length of this breakwater system is 13.52 kilometers (8.4 miles).

How Are Breakwaters Built?

Sometimes we refer to breakwaters as rubble-mound structures. The term is meant to imply that such a structure is built of individual rocks or (if rock isn't available or is too expensive) concrete units. Rocks or concrete units of various sizes are placed, piece by piece or unit by unit, until the desired cross-section is achieved. Figure 27—a schematic drawing of the cross-section of a rubble-mound structure—shows a core composed of small material with two layers of larger material protecting it: a sub-layer and an armor layer. A very large breakwater may have more than one sub-layer.

Smaller rocks are used in the core of a breakwater than in the armor layer and the sub-layer so that the core will be relatively impervious to waves. It is the core that does most of the work of blocking wave activity, and the core is the most important part of the breakwater as far as reflection and transmission of wave energy are concerned. A breakwater that perfectly blocked offshore waves from entering a harbor would have a transmission of 0 percent and a reflectivity of 100 percent.

Waves would wash the core of a breakwater away if it weren't protected by an armor layer and a sub-layer made of larger materials. For economic reasons, a core usually isn't built any higher than the crests of the maximum waves expected at the site and oft times at a lower elevation. If you think of a breakwater's cross-section as

Figure 29 The breakwater at Kahalui, Maui, showing tribar and dolosse armor units and the ribbed concrete cap tying together the seaward and shoreward faces of the structure. (source: author)

a trapezoid (a truncated triangle), an increase in height means adding material to the sides of the trapezoid. This is generally uneconomical. Therefore, because of the lower crest of the core most breakwaters are somewhat transparent to waves and especially to long waves. In other words, some wave energy travels through the spaces between the sub-layers and the armor layer and over the crest of the core, then into the harbor behind the breakwater. Thus, along with the wave energy that travels through the breakwater's harbor entrance, there may still be excessive wave energy within the protected harbor. However, there may be an environmental benefit associated with this wave energy transmission through the structure. The circulation generated by waves and the tides within the harbor or bay is important to maintain the water quality, and the transmission of waves and tide through the breakwater can improve the quality of the water in the harbor by enhancing internal harbor circulation.

The largest pieces of rock in a breakwater are used in the armor layer, the purpose of which is to resist movement by the largest waves that are expected at the site. Aside from the size of the individual rock pieces, the integrity of the armor layer also depends on the density of the rock, on how well the individual pieces interlock, and on the cross-sectional shape of the overall structure.

Let's look at one rock nesting on adjacent rocks and resting on the surface of a breakwater, as illustrated in figure 28. After a wave strikes the breakwater, a "run-up tongue" is formed as the wave travels up the breakwater's seaward face. The velocities and accelerations of the water particles in the run-up tongue may dislodge the rock. The direction of each of the forces is shown by an arrow. There are two forces operating on the rock that act in the direction of the up-slope motion of the run-up tongue. Similar to the forces on a moored ship, the drag force (F_D) is a function of the velocity of the flow, the viscosity of the seawater, and the shape and roughness of a rock piece. The inertia force (F_I) is caused by the fluid accelerating around the rock. If the velocity were steady (an acceleration of zero) the inertia force would be zero. Another fluid force acting on the rock in a direction perpendicular to the slope is the lift force (F_L) caused by the velocity and the shape of the exposed rock. The lift force—shown in figure 28 by the arrow pointing away from and perpendicular to the slope—is analogous to the lift force that acts on the wings of an airplane. The last force involved in the incipient motion of this rock is its submerged weight. (The submerged weight of the rock is equal to its weight in air subtracted by its buoyant force.) The submerged weight, which acts vertically downward, can be broken up into two components, one of which acts in a direction that is perpendicular to the slope and one of which acts in a direction parallel to the sloping face and directed down the slope. (Think of these two components as the two perpendicular legs of a right triangle, and the submerged weight as the hypotenuse.) Thus, the two components of the submerged weight of the rock when the flow is up the slope are opposed to the wave-induced forces: the component parallel to the slope resists the drag force and the inertial force, and the component of the submerged weight perpendicular to the slope resists the lift force. The balance of these forces determines whether the rock will be stable or whether it will move. When there is a down-rush of water on the seaward face of the structure, the drag force is in the direction of gravity enhancing the probability of dislodging armor units if the velocity of the down-rush is sufficient. However, observations in the laboratory generally indicate that armor units are plucked from the face of the structure by wave run-up and then travel downslope during the wave down-rush. In essence this force balance defines the stability of the breakwater.

Since the sole purpose of the armor layer is to protect the core, it is important that the individual pieces used in the core don't wash out through the spaces between the larger pieces that make up the armor layer. That is the function of the sub-layer (or sub-layers). Several sub-layers may be used so the core material doesn't pass through the spaces between pieces of the sub-layer and so that the pieces of rock in one sub-layer don't pass through the spaces between the large material in sub-layer above it or in the armor layer. In a sense, these layers form an inverted

Before earthquake occurs

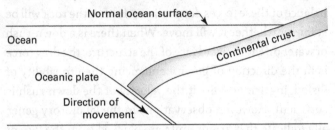

Earthquake occurs

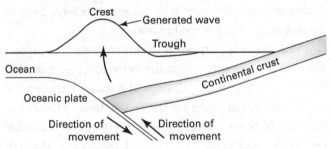

Figure 30 A simplified schematic diagram of a subduction-zone earthquake.

filter, with the size of the rock increasing with distance outward from the core.

Formulas have been developed for designers of breakwaters to use in determining how heavy the individual rocks or the individual concrete units in them will have to be to resist storm waves with various characteristics. The variables in these formulas include the height of the waves, the density of the rocks or units, the angle of the face of the structure, the acceleration of gravity, the density of the fluid (usually sea water), and a coefficient that takes into account the shape of the rocks or concrete units and how they pack together. Different formulas have been developed over the years, but the parameters used are about the same in all of them. According to these formulas (which were derived through consideration of the mechanics of the wave-structure interaction and from the results of laboratory tests of scale models in wave tanks), the weight of the armor unit is directly proportional to the cube of the wave height. This means that if the wave height is doubled, the weight of the unit required to resist the waves increases by a factor of 8. (For a more extensive discussion, see Raichlen 1975.)

Since most large breakwaters are extremely expensive (costs may vary from $20,000 to $80,000 per meter), laboratory tests usually are done before a final design is chosen. A formula such as that mentioned above is used to calculate the approximate size of the armor. After a preliminary design has been created, a scale model of the breakwater (or a section of it) is built. The model is exposed to waves modeled after those expected at the site. If the model is of a small section of the breakwater, it is tested in a wave tank; if it is of the complete breakwater, it is tested in a large wave basin, with the bottom of the ocean at the location of the breakwater modeled accurately to scale. The former is called a *two-dimensional model*; the latter is referred to as a *three-dimensional model*. The core, the sublayer(s), and the armor layer are modeled as accurately as is possible, down to the size, the shape, and the density of the individuals units. To reduce the effects of scale on the results, the model is built as large as is possible. (For more details of model testing, see Hughes 1993.)

In 1981, one of the two breakwaters built to protect the cooling-water intake of the Diablo Canyon nuclear power plant in California was damaged by waves from an extreme storm. To investigate the wave damage and determine how to reconstruct portions of the breakwater, a large three-dimensional scale model of the structure was built. (See Lillevang et al. 1984.) The bottom of the laboratory basin was molded to accurately depict the actual bathymetry, and the modeled breakwaters were exposed to waves corresponding to the wave spectrum from the damaging storm. The accuracy of the laboratory model of the breakwaters and the testing procedure was confirmed when the same damage and the progression of the damage that was observed in the actual structure was seen in the model. This allowed the engineer to have confidence that a design to repair the breakwater based on laboratory tests would result in long-term protection for the plant. The design changes to the breakwater that resulted from the laboratory tests were implemented and appear to be successful to this day.

The shape of the rocks used in the armor layer of a breakwater and the placement of each rock are nearly as important as the size of the rocks. To achieve good packing, very blocky rocks or rounded boulders aren't used. The engineer wants each rock in the armor layer to be longer in one dimension than in the other two dimensions. Thus, each rock piece is placed with its long axis perpendicular to the face of the structure. If this is done consistently, the armor layer will better resist incoming waves. If its isn't done consistently, the armor layer is more likely to fail.

The care that is taken in breakwater construction is exemplified by the outer breakwater of Dana Point Harbor, a small boat harbor, near Los Angeles, built in the late 1960s. This breakwater was to be constructed of rock, and the proposed cross-section was tested in a large wave tank located near Washington, DC. The wave tank was large enough that a 1/5-scale model could be built. A crane operator from the firm that was to construct the breakwater was flown to Washington to build the model using a small-scale crane and the same techniques he would use at Dana

Point. This procedure resulted in a strong structure that has resisted storms for 50 years except for some damage in 1983. (The 1983 storm caused damage to the structure that was repaired in 1984.)

Such care in construction generally wasn't taken years ago, when it wasn't understood that the long-term stability of a rubble-mound structure could be improved by good engineering. A breakwater built in the early 1930s seaward of the amusement pier at Santa Monica was partially destroyed later in the same decade by waves generated by a rare hurricane. That breakwater incurred further damage from storm waves during the winter of 1983; the southern one-third of the breakwater was significantly damaged, and as a result the seaward one-third of the Santa Monica pier was destroyed. For economic reasons, the breakwater was never rebuilt.

In many locations, large rocks aren't available, or perhaps it isn't economical to use large rocks in the construction of the armor layer of a breakwaters or a jetty. In these cases artificial armor units constructed of concrete have been used instead of rocks in the outer armor layers. Sometimes these units are made of un-reinforced concrete; sometimes steel-reinforced concrete is used. Many differently shaped units have been developed over the years. All of them have been designed to interlock like rocks so that the armor layer will be flexible yet strong. The flexibility of the breakwater is important so if some settling occurs the breakwater will still remain intact.

Opinions vary as to whether concrete units used in armor layers should be reinforced with steel bars. Concrete is considered to be a good material to use when it is likely to be exposed to compressive forces. An upright cylinder of concrete to which an increasing load is applied will support the load until the force becomes great enough to rupture the concrete. However, if we were to apply loads to each end of the cylinder that acted in a direction to *pull* the cylinder apart (tension) the concrete would crack at a much lower applied force than when the cylinder was in compression. It is when the concrete cracks in tension that the reinforcing rods buried in the structure take over and prevent total failure. In the ocean, if a unit becomes cracked—even if it is apparently intact—sea water will penetrate and corrode the reinforcing steel. Nevertheless, steel-reinforced concrete is generally used, if only as a secondary line of protection (both when units are placed in a breakwater and while they are being transported from the casting location to the construction site).

Some concrete armor units are better placed uniformly in a layer, interlocking with their neighbors; others, with different shapes, provide the best protection when placed more randomly. Just as designers try to get good packing with rocks, they try to do the same with concrete units. Some individual concrete units that have been used in breakwater construction have weighed more than 50 tons (100,000 pounds). The concrete armor units that were used in the outer layer of breakwaters protecting the cooling water intake structure at the Diablo Canyon Nuclear Power Plant weighed 21.5 tons (43,000 pounds) each.

The photograph presented here as figure 29 shows a portion of the breakwater at Kahalui, Maui. Two types of concrete armor units can be seen: tribars and dolosses. (A tribar consists of three cylinders connected by a yoke; a dolosse is like the letter H with one leg rotated at 90° to the other.) These units interlock on the front and rear faces of the structure, and a ribbed concrete breakwater cap ties the seaward and landward faces of the structure together. The breakwater at Kahalui, built in 1900, has been repaired several times. The size of the units can be appreciated by comparing them to the people walking on the structure. The concrete armor units that have been used to repair it vary in weight from 6.5 tons to 50 tons.

After construction, a breakwater can settle to an extent, becoming more stable than it was immediately after construction. However, when concrete armor units are used, it is desirable to limit the movement of the individual units to prevent their breakage.

Some breakwaters have been built under the assumption that their seaward faces would deform under wave attack. Such breakwaters are called *berm breakwaters*. The rocks used in the armor layer of a berm breakwater can be

Figure 31 A view of a ship brought onshore by the tsunami of March 11, 2011 and local damage. (source: H. Yeh)

smaller than the size required for a stable structure that retains its original shape, so a berm breakwater is less costly to build. A berm breakwater's face is originally plane. With exposure to waves, rock is removed from the region roughly above still water level and deposited down the slope under water. After some duration of wave exposure, the breakwater face is deformed in cross-section. The face of the structure attains equilibrium with the waves much as a beach does. In essence, the breakwater becomes dynamically stable. The design of a berm breakwater requires considerable testing to be ensured that it will not fail.

When waves strike a breakwater, they will run up its face in the same way that waves run up the face of a beach after they break and collapse. Waves generated by extreme storms can overtop a breakwater and run down its harbor face. Many people think that a structure is safe simply because it is massive, and sometimes people venture out onto a breakwater and are exposed to large waves that occur in a group of smaller waves. (Recall the Rayleigh distribution of wave heights in a train of waves.) Thus, never walk out on a coastal structure during a storm. You never know when a large wave might sweep over the structure.

If a breakwater is likely to be overtopped by extreme waves, attention must be paid to the design of its back face (or the harbor side) as well as to its seaward face. Remember, from the paragraph that discussed the incipient motion of a rock nested on the face of a structure composed of similar rocks, that one force that resisted the incipient motion of the rocks caused by fluid forces during the run-up of a wave on the seaward face of a breakwater was due to gravity. However, if the waves overtop the breakwater and run down the harbor face of the structure, the gravity force now acts in the direction of the drag force and the inertia force. This is why, in the design of a breakwater, the harbor face of the structure must be considered as well as the seaward face. What usually minimizes the fluid forces acting on the harbor face of the structure is that, owing to the roughness and the porosity of the seaward face of the breakwater, the overtopping volume is reduced compared to the volume in the run-up tongue on the seaward face.

Also, although a breakwater is constructed to create a quiet area behind it, when a breakwater is overtopped a harbor's tranquility may be significantly reduced. In addition, owing to the constricted entrance to a harbor, overtopping can increase the water level in a protected harbor, an effect referred to as *ponding*.

The run-up of waves on the seaward face of a breakwater depends on both the angle of the front face of the structure and the steepness (ratio of height to length) of the waves offshore in deep water. For example, for a smooth and impervious slope with an angle of about 11° to the horizontal (a slope of about one unit vertical to five units horizontal), the run-up can vary from 0.7 to 4.0 times the deep-water wave height, depending on the steepness of the waves offshore. When the slope is composed of rock the surface becomes rough and the ratio of the run-up to the height of the offshore deep-water waves may be reduced to about one or less.

Even with the reduced run-up on a rubble slope, significant problems can occur. For example, a small artificial island constructed off the shore of Southern California for the purpose of oil recovery and protected by rock revetments on all sides was severally damaged by overtopping waves generated by a large storm. It was essentially "wiped clean" by overtopping waves destroying the drilling and storage structures on the island. Thus, run-up and overtopping can significantly affect the viability of offshore structures as well as those built at the shore.

EXTREME WAVES

Hurricanes (Typhoons, Cyclones)

The extreme wind events that Americans call hurricanes are referred to as typhoons in Asia and as cyclones in Australia. Only the names differ. In this section, these storms will be referred to as hurricanes, and we will try to understand some of the major characteristics of the wind fields that constitute them and the waves that result. Aside from waves, another phenomenon that can be devastating during a hurricane is storm surge, an increase in the water surface elevation at the coast by tens of feet.

Hurricane Katrina, a Category 3 storm that came ashore late in August 2005, was one of the five deadliest hurricanes in American history. It was the sixth strongest of recorded Atlantic storms. There were 1,836 deaths caused by winds and flooding, and the property damage was estimated to be $100 billion. Severe coastal destruction—much of it due to storm surge—extended from Florida to Texas. New Orleans was hit especially hard. Part of its levee system failed catastrophically, and 80 percent of the city experienced flooding.

What Is a Hurricane?

A typical tropical hurricane, when viewed from above, is seen as spiral cloud bands with a central open region called the eye. The Coriolis effect spins the winds (counterclockwise in the northern hemisphere and clockwise in the southern hemisphere) and affects the overall trajectory of the storm. The overall diameter of a typical hurricane is a few hundred kilometers. This spiraling wind (or vortex) has a pressure that varies from a little more than 1,000 millibars at its outer limit to 85–95 percent of that at its center. The pressure at the center of the hurricane is often less than 960 millibars; one of the lowest central pressures ever recorded was 870 millibars. (A pressure of one atmosphere is equivalent to about 1,014 millibars.) The winds spiral inward toward the center. The winds in the eye of the storm are substantially less than those in the outer reaches of the storm—perhaps on the order of 10 kilometers (about 7 miles) per hour. The wind speed varies across the storm from a low in the eye to a maximum at a radial distance from the eye that is a function of the storm intensity. The wind speed then decreases to the speed of the ambient winds outside of the hurricane's influence.

The spiral cloud bands have strong upward velocities separated by regions of downward velocities with velocities at the sea surface directed toward the center. The wind at the sea surface derives its energy from the sea, and the warmth of the water is most important. A tropical hurricane can be viewed as a giant vertical heat engine. Water that has evaporated from the ocean surface travels upward, releasing heat as the water vapor condenses. This released heat drives the destructive winds. Hence, the generation of hurricanes occurs primarily in the tropics, where the sea surface temperature is higher than about 27°C (81°F). Two-thirds of hurricanes form at latitudes between 10° and 20° north and south. The decrease in hurricane generation north and south of this region is due to the less frequent occurrence of high sea surface temperatures.

A hurricane is a dynamic storm that moves at varying speeds across the ocean. Aside from the damage caused by the wind itself, the high wind speeds contribute to coastal hazards by generating large waves and through the action of the winds on the water surface, where they create a shear force that can increase or decrease the water level significantly depending on whether the wind is directed onshore or offshore. For a rotating wind field moving along a path toward land, in the northern hemisphere, winds to the right (as seen from the eye and looking along the path of travel) are directed toward land. Conversely, to the left the winds are in the offshore direction.

The wind at the sea surface derives its energy from the sea, and the warmth of the water is most important. A tropical hurricane can be viewed as a giant vertical heat engine.

Water that has evaporated from the ocean surface travels upward, releasing heat as the water vapor condenses. This released heat drives the destructive winds.

What Is a Storm Surge?

Let's look at the balance of forces on a slice of fluid extending from the surface to the bottom that is aligned in the wind direction and is sheared at the water surface by the wind. Owing to the surface shearing force, the water surface at the downwind end of the slice rises and the upwind end falls. From a balance of forces in the direction perpendicular to the end of the slice, the water surface slope can be shown to be a function of the slope of the sea bottom in the wind direction, the depth, and the wind velocity. (The shear force on the surface and on the sloping bottom are balanced by the gradient of the hydrostatic pressure forces on the ends of the slice.) For the same wind velocity, when the bottom slope is very small, as it is near Galveston, the tilt of the water surface (the storm surge) is much greater than it would be if the same hypothetical storm were to occur near Santa Monica, where the bottom slope is much steeper. To those familiar with the very small offshore slopes at Galveston, the large overtopping of the coast that occurred during the 1900 hurricane isn't surprising.

In addition to the effect of the shearing force of the large wind velocities acting on the sea surface, we have to realize that the overall air pressure on the water surface can cause changes in the water level. Suppose you hold the nozzle of a vacuum cleaner a short distance above the water surface in the center of a bathtub. If you observe the water surface carefully, after you turn on the vacuum cleaner you will see the water level under the nozzle rise slightly. (Be careful, and don't drop the vacuum cleaner into the tub.) Because the tub is a closed basin, the water level away from the nozzle will have to fall slightly. The same thing happens in the eye of a hurricane, where the barometric pressure is lower than the pressure some distance from the eye. Normal air pressure, undisturbed by a hurricane, is about 1,014 millibars (about 14.5 pounds per square inch). Measurements have shown that the pressure in the eye of a hurricane can range from about 960 millibars (about 5 percent less than normal) down to about 870 millibars (about 14 percent less than normal). This decrease in pressure can cause the water surface in the eye relative to the water level outside of the storm area to rise from about 0.5 meter (1.6 feet) to 1.4 meters (4.6 feet). That may seem to be a large elevation, but like many things in the world it is only relative. The increase elevation is usually small relative to the piling up of water at the coast caused by the shearing action of the wind.

If the storm causes strong currents to flow along the shore the Coriolis acceleration can add or subtract from the storm surge, depending on the direction of the current. Associated with wave breaking (discussed in chapter 1) is a "super-elevation" of the water surface called *setup*. This effect can be important with regard to hurricane-generated waves, since the high winds related to hurricanes can generate large waves at the coast. The increase in the storm surge elevation caused by setup can be as much as 20 percent of the breaking waves' height.

Much of the damage caused by Hurricane Katrina was attributable to the storm surge along the Gulf Coast. The surge attained a height of 7–10 meters (23–33 feet) along a 60-kilometer (37-mile) stretch of the coast. In some locations, flooding extended 20 kilometers (13 miles) inland. Surge heights on Lake Pontchartrain ranged from 3 to 4 meters (10 to 13 feet) and contributed to levee failures that resulted in the flooding of New Orleans.

Hurricane Hugo, a Category 4 storm that struck the South Carolina coastline in September of 1989, caused $4 billion worth of damage and 39 deaths. The path of the storm was somewhat erratic, but eventually the eye made landfall near Charleston. The speed of the maximum sustained winds was a startling 230 kilometers (140 miles) per hour, and the central pressure was 944 millibars. The rise in the water surface caused by the lower air pressure in the eye of the storm was about 0.7 meter (2.3 feet). A hurricane is a moving pressure field, and Hugo traveled at between 32 and 40 kilometers (20 and 25 miles) per hour. The height of the surge at Charleston was nearly 4 meters (12.9 feet). Remember that the wind direction of tropical cyclones in the northern hemisphere is counterclockwise, so the maximum winds, waves, and surge are to the right of the eye (looking in the direction of storm travel). Hugo covered a huge area—about the size of the

Much of damage caused by Hurricane Katrina was attributable to the storm surge along the Gulf Coast. The surge attained a height of 7–10 meters (23–33 feet) along a 60-kilometer (37-mile) stretch of the coast. In some locations, flooding extended 20 kilometers (13 miles) inland.

state of Georgia. About 32 kilometers (20 miles) north of the entrance to Charleston Harbor (to the right of the eye of the hurricane), the storm surge was nearly twice that at Charleston—6 meters, or 20.3 feet. Fortunately, that area was lightly populated. With its somewhat erratic path, Hurricane Hugo could have done much more damage to Charleston if the eye had struck the shoreline only 20 miles (32 kilometers) farther south.

Hurricane-Generated Waves

The high winds also generate large waves on top of the storm surge. The action of these waves may be very destructive to the beaches and to buildings located near the coast. In essence, the increase in the water level at the coast moves the shoreline inland. This allows waves to attack regions that previously were safe from naturally occurring storm waves. For example, Hurricane Hugo caused erosion to beaches along a 240-kilometer (150-mile) swath extending northward into North Carolina. This created a serious problem for owners of beachfront homes, since it removed the protection from waves that the beach (and perhaps dunes) in front of their property had provided in the past.

The waves generated by a hurricane differ from those generated by normal wind conditions in several ways. The speeds of the winds in a hurricane aren't constant, so the wave characteristics vary. The winds in a hurricane have a circular path. The hurricane moves over waves generated at various angles of direction to the storm path (Ippen 1966). To the right of the eye (in the northern hemisphere), the winds are enhanced by the forward speed of the hurricane increasing the wind speed from simply that of the rotating wind system. Because the storm surge increases the water's depth near the shore, the waves generated by the wind can now propagate on deeper water, and larger waves and wave breaking can occur nearer to the shore than would occur without the hurricane. Of course, this effect is exacerbated if the hurricane arrives at high astronomical tide.

The U.S. Army Corps of Engineers has developed methods for predicting hurricane-induced waves in deep water. Some of these results are presented here in table 2. The methods used to predict the wave height and wave period ranges shown in the table rely heavily on observed data, and they are simply a guide to what might be expected to occur.

Tsunamis

Some History

On December 24, 2004, an earthquake with a moment magnitude that exceeded 9 occurred off the shore of the Indonesian island of Sumatra and caused a loss of between

Table 2   Hurricane categories and wave characteristics. (source: 2001 U.S. Army Corps of Engineers Coastal Engineering Manual)

Category (Saffir-Simpson hurricane scale)	Wind speed (kilometers per hour)	Wind speed (miles per hour)	Wave height range (meters)	Wave period range (seconds)
1	119–153	74–95	4–8	7–11
2	154–177	96–110	6–10	9–12
3	178–209	111–130	10–14	12–15
4	210–249	131–155	10–14	12–15
5	>249	>155	12–17	13–17

200,000 and 300,000 lives. Its effects were felt as far away as Somalia, on the Horn of Africa. (The moment magnitude is used now by seismologists to measure the size of an earthquake in terms of the energy released. It replaces the Richter scale. See Hanks and Kanamori 1979.) In Thailand, the first indication of a tsunami was the recession of the sea that exposed the bottom near the shore. Many beachgoers were overpowered by the wave that then came roaring in.

A catastrophic earthquake and tsunami occurred near the northeast coast of the Japanese island of Honshu on March 11, 2011. The earthquake's moment magnitude was 9.0, and the shaking lasted about 5 minutes. This was the most powerful earthquake ever to hit Japan, although one in the year 869 may have caused similar damage. It was generated by the Pacific Plate subducting under the plate that lay beneath northern Honshu. The area of significant fault displacement had a length of about 250 kilometers (155 miles), a width of about 100 kilometers (60 miles), and bottom displacements of about 10 meters (33 feet). Subsidence of the coastal region exacerbated local flooding. The tsunami that this earthquake generated, with an onshore height of 15 meters (50 feet) in some areas, swept away coastal villages from the city of Sendai northward. That beautiful part of the Japanese coast had many fishing villages in valleys extending landward from the coast. In one location, the tsunami swept nearly 10 kilometers (6 miles) up a valley, completely erasing a village. More than 16,000 people were killed with about 4000 missing on Honshu. Fishing boats, cars, and debris from destroyed houses were carried inland, causing unprecedented damage. In one town, people seeking refuge on the roof of a three-story concrete building 13 meters (42.6 feet) tall were swept away. In some areas, the tsunami came ashore as an unbroken wave, acting more like a fast-rising tide; but in other locations it acted like a huge broken wave or a surge.

Tsunamis have occurred in the Bay of Bengal, in the Caribbean Sea, in the Celebes Sea, in the Java Sea, in the Mediterranean Sea, in the South China Sea. and in the Atlantic Ocean. (On November 1, 1755, a tsunami caused

The area of fault displacement had a length of about 250 kilometers (155 miles), a width of about 100 kilometers (60 miles), and maximum bottom displacements of about 10 meters (33 feet).

great destruction in Lisbon, Portugal.) However, the countries most at risk from tsunamis have coastlines on the Pacific Ocean. Large earthquakes occur along the "Ring of Fire" (the arc of volcanoes around the Pacific Rim), some of them undersea quakes capable of generating destructive tsunamis. It has been estimated that 80 percent of major tsunamis originate in that region.

Although a tsunami is generated by a single sudden event, a tsunami isn't a single wave. It consists of a series of waves, some of them larger than the one that is the first to reach the shore. These waves may arrive tens of minutes apart, and the wave activity may go on for hours. Depending on the characteristics of the waves and the bathymetry of the coast, the waves may arrive as breaking waves or as a fast-rising tide.

For many years, earthquake-generated sea waves were called "tidal waves." That term is still in use, but it is confusing because they have nothing to do with the astronomical tides or with storm tides. For that reason, the Japanese term *tsunami* was adopted. Its literal meaning is "harbor wave." This term derives from the past great destruction in harbors around the coast of Japan.

Tsunamis are generated by impulsive boundary motions. Among the mechanisms that may cause tsunamis are earthquakes (tectonic events), volcanic eruptions, above-water (sub-aerial) and/or below-water (sub-marine) landslides, and even explosions. (In the mid 1950s, some

The definition of the word *tsunami* does not refer to a wave generated by an earthquake. The literal translation is "harbor wave." This term derives from the great destruction that has occurred in past years, including the 2011 event, in harbors around the coast of the Japanese Islands.

research was conducted to explore the possibility of using the waves generated by an underwater hydrogen-bomb explosion as a weapon.) All these mechanisms have one thing in common: in some manner, energy is put into the water column by a short-time disturbance at either the solid-water boundary or the air-water boundary.

Earthquake-Generated Sea Waves

Except for several important exceptions that will be discussed later, an earthquake must occur underwater, not on land, to generate sea waves. For purposes of the present discussion, let's simply consider the movement of the solid-water boundary at the ocean bottom. To generate a wave at the sea surface during an earthquake, the bottom has to physically displace a mass of water. The bottom motion that causes this displacement is vertical and referred to as a *dip-slip boundary motion*. One side of the earthquake fault moves upward while the other moves downward. If one side of a fault slides relative to the other without a significant vertical displacement (a *strike-slip*), the disturbance to the water surface will be minimal. However, in some cases the ground motions at the ends of a strike-slip fault may generate tsunamis that affect local regions. (A tectonic motion will usually involve both dip-slip and strike-slip motions.) The vertical displacement of the bottom that generates an unusually powerful tsunami may be on the order of many meters.

To understand how motion of the bottom can generate a wave, sit in a bathtub and place one hand flat on the bottom of the tub. If you slide the hand along the bottom, you will not see much of a disturbance to the water surface. However, if you impulsively (that is, suddenly) lift the hand a short distance from the bottom and the water is fairly shallow, you probably will see a disturbance on the surface—a small tsunami. Now repeat this little experiment by first putting your open hand a few centimeters above the bottom and then impulsively moving it downward. You may see a depression at the water surface, and then see waves moving outward from that disturbance. If the water in the tub is deep, the wave generated will not be as large as it will be if the water is shallow. Thus, the relative size of the disturbance (in the case of the example of the tub, the size of your hand relative to the depth) appears to be important in this method of wave generation. The ratio of the size of the bottom disturbance to the depth is indeed important to the generation of a tsunami in an ocean. However, it is necessary to be a bit more specific. It's the ratio of the size of the bottom disturbance in the direction the wave will travel away from the source to the depth that is important.

Now climb back into the bathtub and repeat the experiment, but now move your hand upward slowly. The disturbance on the water surface (if there is any) will be small relative to the one that occurred when you moved your hand very rapidly the same distance. Now an additional factor enters into the question of wave generation by bottom motions: the speed of the bottom motion. In fact it isn't the absolute speed of the bottom motion that is important, but the ratio of the speed of the bottom motion to the speed of the wave that propagates from the generation region. The wave generated will be larger if the speed of the bottom movement is greater than the speed of the waves that move from the generation region.

Thus, in order for a large tsunami to be generated, the bottom must be displaced upward or downward, the movement of the bottom must be rapid, and the "footprint" of the disturbance must be large relative to the depth. Thus, the earthquake has to be large in magnitude and has to have a relatively shallow hypocenter. The Japanese investigator K. Iida (1961) estimated that the earthquake has to be more powerful than about 7 on the Richter scale and that the hypocenter has to be less than about 30–40 kilometers (18.6–25 miles) below the ocean bottom to generate a destructive tsunami. Recent evidence suggests that these constraints may be somewhat conservative. The current thinking is that the magnitude of the earthquake must be at least 6 and the depth to the hypocenter should be less than about 30 kilometers (18 miles). Local coastal conditions also play a significant role in determining the destructive nature of tsunamis, as they do for normally occurring storm waves.

A tsunami's wavelength in the deep ocean may be on the order of several hundred kilometers, with a wave height of only a fraction of a meter. Considering that the ocean's depth is 3 to 4 kilometers, tsunamis are shallow-water waves. Owing to processes discussed in chapter 1, a tsunami can become many meters high at the coast. Owing to its large wave length as it moves on shore, an immense volume of water flows over the land.

Modeling the undersea bottom movement associated with a large earthquake as a simple block movement upward or downward is simplistic; what actually happens is much more complex. For example, the 2004 earthquake in Sumatra and the 2011 earthquake in Japan were subduction-zone earthquakes—that is, one of the Earth's tectonic plates was moving under another. The bottom movement was upward to one side of the fault and downward to the other side. Figure 30 is a simplified diagram of the bottom motion in a subduction-zone earthquake. The downward motion of the subducting plate drags the upper plate downward as shown in the upper portion of the figure. When the upper plate can no longer resist the downward plate's motion, it rebounds, flinging a huge volume of water upward. It is this displaced column of water that produces the tsunami, from which a series of waves propagates both shoreward and seaward. The idealized plate motions are indicated in the figure by arrows, and the formation of the tsunami by the "hump" of water on the sea surface above the plate intersection.

Figure 32   A seawall built to protect the Japanese fishing village of Taro from tsunamis. The houses in the foreground are outside the seawall. (source: author)

The fault motion during an earthquake is like a "ripping" movement. The disturbance starts in one region, then proceeds along the fault at the fracture speed of rock (about 3.2 kilometers or 2 miles per second). In Sumatra the faulting took place over a distance of more than 1,000 kilometers (600 miles), and the ground shook for several minutes. In the case of the 2011 Japanese earthquake the rupture length was much shorter. It was about 250 kilometers—about the distance from Los Angeles to San Diego.

An extremely simple analogy to the fault motion and generation process in an earthquake like the 2004 Sumatra event could be the keyboard of a piano. Suppose you could move adjacent piano keys on the keyboard upward, one at a time, in quick succession. In that analogy, each key represents a section of the fault, the length of the keyboard represents the total length of the faulting, and the upward motion of each key simulates the impulsive movement of a section of the fault. If we now translate this idea of incremental fault motion propagation to a progressive process of wave generation, a water wave will be produced as each section of the fault moves, each section contributing both to the local and to distant tsunami effects. (This is certainly an extremely simplified example of a very complicated process.)

An earthquake similar in magnitude to the 2004 Sumatra quake or the 2011 quake in Japan appears to have occurred in 1700 off the shore of the state of Washington, in the northwestern US. (This earthquake probably occurred along the Cascadia fault that has the potential for a major subduction zone earthquake.) Geologic investigations found sand layers and buried trees along the coast of Washington that date from about 1700 that may be related to waves larger than those generated by an offshore storm. The event must have carried a significant amount of sand landward from the offshore regions. Additional confirmation of this event was discovered by researchers in Japan who, while searching their archives, found a reference to a tsunami on January 26, 1700 and a description of the damage it caused in Japan. And Native American lore in Washington alludes to a wave washing over the Olympic peninsula (Heaton et al. 1985; Atwater et al. 2005).

In the 2011 earthquake and tsunami, in many coastal towns both small boats and large ships became floating missiles that were carried far inland by the inundating waves. Figure 31 shows a large ship brought onshore by the tsunami resting on the debris of a destroyed village. The earthquake, which occurred about 70 kilometers (43 miles) offshore at a depth of about 32 kilometers (20 miles), was one of the most powerful ever to have struck Japan and the fifth-largest recorded anywhere in the world. The tsunami reached the coast about 40 minutes after the earthquake was felt. The time between the earthquake and the appearance of the first wave gave residents some time to evacuate, but the height of the tsunami and the distance

The Tohoku earthquake occurred about 70 kilometers (43 miles) offshore at a depth of about 32 kilometers (20 miles). It was one of the most powerful earthquakes to have struck Japan and the fifth largest recorded worldwide. The tsunami reached the coast about 40 minutes after the earthquake was felt.

inland it traveled (a number of kilometers in some locations), the speed of the wave traveling onshore, and the debris entrained made it difficult for many to escape. Education about the danger of tsunamis and clear designation of evacuation routes saved many lives.

Japan has always been very conscious of the risk of tsunamis, owing to its proximity to the "Ring of Fire" and its associated earthquakes and its direct and unprotected exposure of its east coast to tsunamis generated by distant earthquakes,. The earliest tsunami researchers, tsunami educators, and constructors of tsunami-mitigation structures were Japanese. Thanks to this attention to tsunamis, Japan probably incurred less loss of life from the 2011 tsunami than a similar tsunami would have years ago.

Many times it is reported that the first indication at the shore that a tsunami had occurred is recession of the water, exposing the bottom. Anecdotes from Japan suggest that in past years this led to the loss of many lives in fishing communities when the receding water exposed fish on the bottom and fishermen went to retrieve them not realizing that a tsunami consists of a number of waves.

If during an earthquake the movement of the ocean bottom is of the dipole kind, with one side of the fault moving downward while the other side moves upward, different waves will be generated moving onshore and offshore. If the negative bottom movement faces the coast, the first wave seen probably will be a negative wave with

Figure 33 A portion of the Taro seawall after the tsunami of 2011. (source: D. Kriebel)

an amplitude below the average level of the sea. In other words, the water at the coast will first recede. On the other hand, the amplitude of the generated wave traveling seaward will be positive—that is, its elevation will be above the mean sea surface. This happened in the 2004 Sumatra earthquake and tsunami. The first wave to arrive in Sumatra was a negative wave, with the surface of the sea receding. In Sri Lanka, more than 1,600 kilometers (1,000 miles) to the west, the first wave to strike the coast was a wave of positive elevation. Therefore, if you are at the coast and you feel a strong earthquake, you should run for higher ground whether or not the near-shore waters recede.

Earlier I mentioned that one mistake people living near the coast sometime make is to assume that a tsunami is a single wave. As a result of the nearly instantaneous bottom displacement, the column of water above the bottom is moved upward rapidly, producing a disturbance on the sea surface that mimics the bottom movement. This nearly instantaneous surface disturbance breaks down and generates a series of waves, see Hammack 1973. Unfortunately for those living at the coast, the first wave may not be the largest. Later waves generated from this initial disturbance may be even larger than the first wave. Local resonance effects (discussed in chapter 4) may also come into play. For example, for several tsunamis at Crescent City, the major damage to the harbor occurred hours after the first wave struck the coast.

Other Mechanisms of Tsunami Generation

Until now I have talked about the generation of tsunamis by undersea earthquakes. Other generation mechanisms have produced destructive tsunamis, however.

The tsunami generated by the eruption of the Krakatoa volcano in 1883 was observed worldwide. The disturbance at the water surface generated a wave that then traveled for thousands of miles.

Tsunamis also can be generated by landslides—either rock-falls above the water surface (called sub-aerial landslides) or by underwater (sub-marine) landslides occurring near the coast. Either a sub-aerial or a sub-marine landslide can be caused by a nearby earthquake. A sub-aerial landslide occurred in Lituya Bay, in Alaska, on July 9, 1958. Lituya Bay is about 14.5 kilometers (9 miles) long and 3.2 kilometers (2 miles) wide at its widest and has a narrow opening to the sea. The normal tidal range is about 3 meters (9.8 feet). A nearby earthquake with a magnitude of 7.9 on the Richter scale caused a sub-aerial landslide on a slope at one end of the bay that generated a wave that was large enough to run up slopes around the bay to an elevation of 524 meters (1,720 feet)—nearly 40 percent higher than the Empire State Building. That wave toppled trees and stripped the soil down to bedrock around the bay.

Landslides caused by earthquakes can also occur underwater. On July 17, 1998, an earthquake with a moment magnitude of 7.1 occurred about 25 kilometers (16 miles) from the coast near Aitape in Papua New Guinea. Many investigators think the earthquake caused a submarine landslide that, in turn, generated a tsunami that killed about 2,000 people and left 9,500 homeless and 500 missing. A coastal spit about 30 kilometers (19 miles) long, running northwest from Aitape to the village of Sissano, was wiped clean. Survey teams that reached Papua New Guinea after the event observed debris high in remaining trees, which indicated that the wave must have been many meters high as it passed over the spit. The effect was felt a great distance inland.

The Aleutian tsunami of April 1, 1946 caused remarkable run-up and splash-up at coastal cliffs on Unimak Island, Alaska. The Scotch Cap Lighthouse had a foundation 13.7 meters (45 feet) above sea level. The tsunami obliterated the lighthouse and the mast of a radio tower located behind it on a cliff 32.3 meters (106 feet) above the sea. There remains some controversy as to whether this excessive run-up and splash-up was caused directly by the tsunami generated by the tectonic bottom motion or by a wave generated by some other mechanism—perhaps an underwater landslide or both. This tsunami caused 159 deaths in the Hawaiian Islands—about 2,000 miles away.

Protecting People and Property

Entrainment of debris by the onshore run-up of a tsunami was observed in videos taken during the 2004 tsunami in

Sumatra and during the 2011 tsunami in Japan. In both cases, cars, trucks, ships, and debris were swept along by both the incident and the outgoing waves. People who might survive being in a current of water are less likely to survive being in a debris-loaded stream.

The force of a tsunami traveling onshore is considerable. In 1960, for example, parking meters at a lot in Hilo, Hawaii were bent like spaghetti by a tsunami that had been generated in Chile, thousands of miles away.

In addition to the loss of life and property damage caused by the extreme coastal inundation of a tsunami, salt water may kill vegetation and may contaminate the groundwater inhibiting farming for many years.

Anyone attempting to design a tsunami-safe structure should understand the structural forces that tsunamis generate. In 1974 a hotel designed to minimize tsunami damage (the Castle Hilo Hawaiian Hotel) was built on the coast in Hilo, Hawaii. Hilo had sustained significant damage in 1946 from a tsunami generated in the Aleutian Islands, and again in 1960 from a tsunami generated in Chile. To protect the upper floors, the Castle Hilo Hawaiian Hotel's parking garage was put on the ground level, with the lobby and the dining rooms on the floor above and the guest rooms on higher floors. The rectangular-cross-section supporting columns in the garage were oriented with their smaller dimension perpendicular to the potential tsunami direction to minimize the transmission of force. After the 1960 tsunami the city of Hilo created a park where buildings destroyed by the tsunami had been located. It was assumed, rightly or wrongly, that a future tsunami could not cause more damage than was caused by the 1960 event. This building-free area was a least-cost solution to tsunami protection for the downtown Hilo region.

At susceptible locations along the West Coast of the United States and in Japan, signs direct people to higher ground or to tsunami shelters. In some locations in Japan where there is no high ground nearby, elevated platforms have been built.

In Japan, a variety of engineering projects intended to mitigate tsunami damage have been undertaken, including the construction of large offshore breakwaters and seawalls to protect coastal cities and towns. In 2009, a breakwater 1,950 meters (6,400 feet) long and located in a depth of 63 meters (207 feet) (the largest depth in the world where a breakwater has been built) was completed to protect the port of Kamaishi. Its construction took 30 years and cost about $1.5 billion. It was damaged and overtopped by the 2011 tsunami, and about 1,100 residents of Kamaishi were killed or went missing; However, without the breakwater more lives certainly would have been lost.

Another tsunami-protection structure built in recent years in Japan was a seawall at the fishing village of Taro, in northeast Honshu. The seawall is 10 meters (32.8 feet) high, about the height of a three-story building. Figure 32 shows this seawall before the 2011 tsunami. Large gates spaced along the seawall could be opened to allow people living outside the seawall to come inside it in the event of an approaching tsunami. During the 2011 tsunami, the seawall was overtopped by several meters. Parts of the structure were destroyed (see figure 33), and portions of the village were inundated. There were major losses of life and property.

Generally a probabilistic approach is used to design coastal structures to resist extreme storm waves. In a probabilistic approach, one tries to establish the recurrence interval of a certain event so that the element of acceptable risk can be incorporated into the design. An example of the probabilistic approach is the designing of coastal structures to resist waves generated by a storm of a magnitude expected to occur once in, say, 50 or 100 years. This is difficult to do for tsunamis, since the historical record is skimpy.

Aside from engineered solutions, the major elements of effective tsunami protection are education, warning, and evacuation.

Warning Systems
Along the north shore of the Hawaiian island of Oahu, sirens are mounted on telephone poles along the highway. These sirens sound if a tsunami is expected, and provide notice to evacuate and move inland. (They were sounded

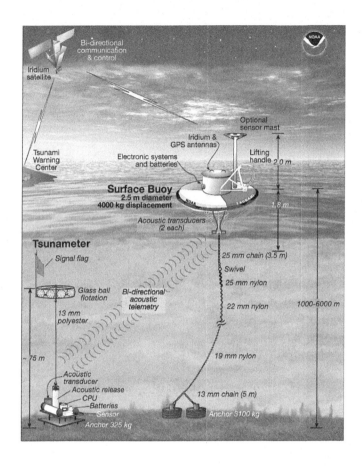

Figure 34 A schematic diagram of the DART II system used by the National Oceanic and Atmospheric Agency to measure tsunamis in the open ocean. (source: NOAA)

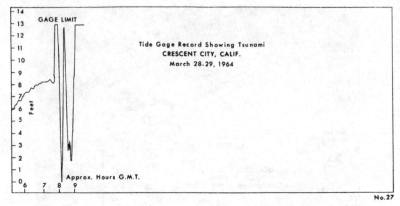

Tide Gage Record Showing Tsunami
CRESCENT CITY, CALIF.
March 28-29, 1964

GAGE LIMIT

Approx. Hours G.M.T.

No.27

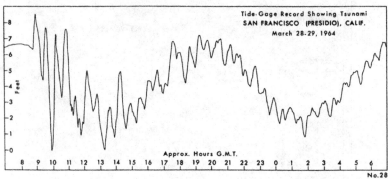

Tide-Gage Record Showing Tsunami
SAN FRANCISCO (PRESIDIO), CALIF.
March 28-29, 1964

Approx. Hours G.M.T.

No.28

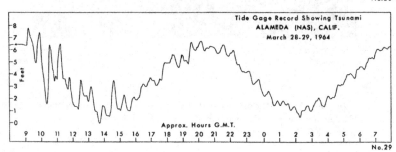

Tide Gage Record Showing Tsunami
ALAMEDA (NAS), CALIF.
March 28-29, 1964

Approx. Hours G.M.T.

No.29

after the 2011 Japanese tsunami and some evacuation took place even though only minor damage was expected.)

Japan was caught off guard, to some extent, by the Chilean earthquake and tsunami of 1960. In one coastal location, a number of people were killed by the waves generated by the earthquake thousands of miles away. There is a monument there to commemorate those victims. In years past monuments were placed at several villages along the coast indicating the extent of the run-up of past tsunamis and admonishing people not to build any closer to the shore. Unfortunately, in many villages that admonition wasn't followed.

The warning system used in the United States at time of the 1960 Chilean event was quite crude. It depended on determining the magnitude of the earthquake and whether tide gauges at selected locations had recorded a wave. The data had to be interpreted before coastal sites could be informed. Since then, a sophisticated tsunami early warning system has been developed by the Pacific Marine Environmental Laboratory (PMEL), a division of the National Oceanic and Atmospheric Agency. Among the system's elements are DART buoys that can measure the height of a tsunami in the deep ocean and transmit the

Figure 35   Tide-gauge records from three locations in Northern California during the Alaskan tsunami of March 28, 1964. (source: Spaeth and Berkman 1967)

data to the PMEL. (The acronym DART stands for Deep-ocean Assessment and Reporting of Tsunamis.) Computer programs developed by the PMEL use the wave measured in deep water by the DART buoys to predict the path of the tsunami across the ocean and the height of the tsunami at specified locations along the coast. This approach has been confirmed over the years by comparing the predicted coastal effects for various tsunamis to coastal observations.

The elements of a DART II buoy are shown schematically in figure 34. The most important element is a pressure transducer placed on the sea bottom at a depth between 1,000 meters (3,280 feet) and 6,000 meters (19,600 feet). The pressure transducer can accurately measure small changes in the elevation of the sea surface. Since a tsunami in the deep ocean is a shallow-water wave, the pressure on the bottom is equal to the depth plus the wave amplitude. The measurements are triggered by an earthquake event. A record of pressures is transmitted acoustically through the ocean depth to a buoy floating on the surface. The signal is then transmitted from the buoy to a satellite and from the satellite to a land-based station, where the record of the water surface is input to complex computer programs that predict wave data for shore locations. This system can supply relatively rapid estimates of tsunami wave heights and arrival times at many coastal locations.

After the Chilean earthquake of February 27, 2010 (a moment magnitude of 8.8), a tsunami warning issued to the Hawaiian Islands caused sirens to sound around Oahu. The warning, perhaps premature, led to extensive evacuation from the low-lying beaches of Oahu. But the PMEL's analysis, the predictions resulting from measurements taken by DART buoys, and the computer modeling suggested that only small effects would be felt on Oahu. That suggestion was ultimately borne out; however, the authorities took a cautious approach.

Deep-ocean measurement of tsunamis was once considered to be an important element in assessing the hazard associated with tsunamis only along the West Coast of the United States and the West Coast of South America. In 2001 three instruments were located near Alaska, two near the coasts of Washington and Oregon, and one near the coast of South America. But in March of 2008, as a result of the 2004 Sumatra earthquake and tsunami, the DART array was expanded to 39 locations in the Pacific Ocean, the Atlantic Ocean, the Gulf of Mexico, and the Caribbean Sea.

For the West Coast and for the Hawaiian Islands, the Pacific Tsunami Warning Centers on Oahu and in Alaska issue warnings of the possibility of a tsunami being generated by a distant earthquake. Warnings urge residents to evacuate low-lying coastal areas. In an area that may be affected, posted warning signs show the direction of a safe evacuation route. Since a tsunami can consist of a number of waves tens of minutes apart, it is important to remain in a shelter until you are informed that the threat of a tsunami has passed.

Figure 35 shows tide-gauge records from the Alaskan tsunami of March 28, 1964 recorded at three locations in California: Crescent City, San Francisco, and Alameda. Crescent City is about 500 kilometers (300 miles) north of San Francisco, and Alameda is on San Francisco Bay. Each of the three records shows the water level as a function of time. The vertical axis shows the level in feet; the horizontal axis shows the elapsed time in hours. The slow up-and-down motion that can be seen in the San Francisco and the Alameda recordings is the astronomical tide. Superposed on this is the recording of the tsunami. At Crescent City, the water surface rose 4.5 feet above the tide level and then fell about 8 feet below the tide level. The limits of the gauge were exceeded, and after about an hour the gauge became inoperable. At the other two locations the height of the tsunami was much less than that observed at Crescent City, and oscillations of the water surface continued for many hours. This "ringing" appears in most coastal tsunami records. The first group of waves is apparently the incident tsunami, and the oscillations of the water surface after the tsunami strikes are waves reflecting in the harbor or bay or even from the coastlines thousands of kilometers away. These surface oscillations do not tell the full story. They are evidence of the coastal water put into motion by the tsunami, and they show that one must be vigilant for hours after the first wave strikes. These currents can play havoc with moored vessels. Even large ships have been pulled from their moorings by tsunami-induced currents.

A tsunami can be generated by a near-shore or distant tectonic event, by a volcanic eruption, or by an above-water or an underwater landslide. In any of these cases, people may have only minutes to evacuate. It is best to follow the Japanese practice: If you are near the coast and you feel shaking due to an earthquake, immediately follow the signs to a safe haven, and in any case heed the advice of the authorities.

I hope to have enhanced your understanding of waves and how they affect us. I have discussed waves ranging from small summer waves to monstrous tsunamis. I have described several issues that bear on the viability of our ports. I've answered questions such as "Why does the Bay of Fundy have immense tide ranges?" I've also touched on a few engineering topics related to how we protect the coast and ourselves from the effect of waves. If you want to study any of these phenomena in more depth, I suggest that you consult the bibliography and the suggested readings.

APPENDIX: EQUATIONS

Newton's Second Law

$$F = m\left(\frac{dV}{dt}\right),$$

where F is the applied force, m is the mass to which the force is applied, and dV/dt is the acceleration that results from the application of the force.

Wave celerity

$$C = L/T,$$

where L is the wave length and T is the wave period. For deep-water waves,

$$C_{\circ} = \frac{g}{2\pi} T,$$

where g is the acceleration of gravity and T is the wave period. For shallow-water waves,

$$C = \sqrt{gh},$$

where h is the depth. For solitary waves,

$$C = \sqrt{g(h+H)}.$$

Wave length

For deep-water waves,

$$L_o = \frac{g}{2\pi} = T^2,$$

For shallow-water waves,

$$L = \sqrt{gh}T.$$

Maximum horizontal water-particle velocity under a wave

For deep-water waves,

$$\frac{U}{C_o} = 2\pi \left(\frac{a_c}{L_o} \right),$$

where u is the maximum horizontal water-particle velocity, a_c is the amplitude of the crest, and L_o is the wave length. For shallow-water waves,

$$\frac{u}{C} = \left(\frac{a_c}{h} \right),$$

where u is the maximum horizontal water-particle velocity, C is the celerity, and h is the depth.

Bottom pressure under a wave

Under the trough,

$$\frac{p}{\gamma} = h - \frac{a_t}{\cosh(2\pi h / L)} > h - a_t \,,$$

where a_t is the amplitude of the trough, p is the pressure, γ is the specific weight of the fluid (= ρg), ρ is the density of the fluid, L is the wave length, and h is the depth. Under the crest,

$$\frac{p}{\gamma} = h + \frac{a_c}{\cosh(2\pi h / L)} < h + a_c \,,$$

where a_c is the amplitude of the crest.

Green's Law (for shallow-water waves)

$$\frac{H_2}{H_1} = \left(\frac{h_1}{h_2}\right)^{1/4} \,,$$

where H_2 is the wave height inshore, h_2 is the depth inshore, H_1 is the wave height offshore, and h_1 is the depth offshore.

Geostrophic wind velocity

$$U_g = \frac{1}{\rho f} \frac{\Delta p}{\Delta n},$$

where ρ is the density of air, $f = 2\omega \sin\phi$ (the Coriolis parameter), ω is Earth's rotational speed (= 0.2625 radian/hour), ϕ is the latitude, Δp is the pressure difference between adjacent isobars, and Δn is the distance between adjacent isobars.

Significant wave height for waves leaving a storm area

$$H_s \approx 0.21 \left(\frac{U^2}{g} \right),$$

where H_s is the significant wave height, U is the wind speed, and g is the acceleration of gravity.

Wave period of peak of spectrum for waves leaving a storm area

$$T_p \approx 7.2 \left(\frac{U}{g} \right),$$

where is T_p is the wave period of the peak of the spectrum.

bar
an offshore sand deposit

bathymetry
the topography of the sea bottom

beach nourishment
placing sand on a beach to rebuild it

berm
the portion of a beach that is inland of the shoreline

centrifugal force
the force that arises in connection with rotation and is directed outward from
the center of rotation

Coriolis effect
the tendency for large masses of fluid to move toward the right in the northern hemisphere (and to the left in the southern hemisphere) relative to an
observer looking toward the equator

cross-shore (offshore/onshore) transport
the movement of beach material in a direction perpendicular to the shoreline

deep-water wave
a wave with the ratio of its depth to its wavelength greater than $1/2$

diurnal tide
one high tide and one low tide per day

drag force
a force imposed on a moving body or a stationary body in a moving fluid; a function of the velocity, viscosity of the fluid, and the body shape and roughness, it acts in a direction opposite to the motion of the body or the fluid

ecliptic plane of Earth
the plane of the Earth's orbit around the sun

fall velocity
the vertical velocity of a particle falling in a fluid

fetch
in a wind field, the distance over which the wind's velocity and direction are relatively constant

foreshore
the inshore region of a beach

geostrophic wind speed
the wind speed that arises from a balance between the pressure gradient force and the Coriolis force in the equation of motion

groins
relatively short structures built in groups, perpendicular to the shoreline, to stabilize the coast

heave
the motion of the center of gravity of a ship in the vertical direction

hydrostatic pressure distribution
the linear distribution of pressure beneath the free surface in a still fluid varying from zero at the free surface to a maximum at the bottom

inertia force
the force acting on a body that results from the acceleration of water particles away from an accelerating body, or the force acting on a body that results from a fluid accelerating past it

isobars
lines of constant pressure in the atmosphere

littoral drift (longshore transport)
the movement of sediment along the coast by waves

moment magnitude
the magnitude scale used by seismologists to indicate the size of an earthquake in terms of the energy released; based on the seismic moment of the earthquake, it supersedes the Richter magnitude scale

natural period
the period of the free oscillations of an elastic system or a body of water

neap tide
the astronomical tide when the moon is in its first quarter or its last quarter

Rayleigh distribution
a particular probability distribution where the mode is less than the mean usually used to describe ocean waves

resonance
a dynamic condition that occurs when an exciting force and the natural response of the system are synchronous

revetment
a coastal protective structure, composed of rock or concrete armor units

Richter scale
a scale used in the past to define the magnitude of an earthquake proportional to the amplitude of earthquake waves measured by a seismograph; superseded by moment magnitude

rogue wave
an extreme wave (usually in deep water) that appears to occur spontaneously and exceeds the significant wave height by more than a factor of 2

seawall
a structure built at the coast to reflect incoming waves so as to protect inland structures

semi-diurnal tide
two high and two low tides per day

shallow-water wave
a wave with a ratio of depth to length less than 1/20

significant wave height
the average of the highest one-third of the wave heights in a group of waves

spectrum
the distribution of a property (for example, energy) with wave frequency or wave period

spring tide
the astronomical tide when the moon is full or new

storm surge
the rise in the water surface at the shore associated with an extreme wind blowing over its surface

surge direction
the motion of the center of gravity of a ship in its longitudinal direction

sway direction
the motion of the center of gravity of a ship in the direction of the beam

swell
the larger-period waves present some distance from a storm region

tidal bore
a wave with an abrupt front face moving upstream in a river, caused by the interaction of the tide with the river flow

tidal range
the distance between low tide and high tide

wave amplitude
the vertical distance from the undisturbed water surface to any point on the wave

wave breaking
the process leading to the destruction of the wave after it reaches its maximum stable height

wave celerity (C)
the velocity of the wave form

wave crest
the distance from the undisturbed water surface to the maximum amplitude of the wave

wave diffraction
the process by which wave energy propagates into regions behind obstacles (islands, breakwaters, headlands, etc.)

wave frequency
$1/T$ (the inverse of wave period)

wave height (*H*)
the distance between a wave's trough and its crest

wave length (*L*)
the distance, at a particular time, between recurrences of like features of a wave, such as the crest

wave period (*T*)
the time, at a particular location, between recurrences of like features of a wave, such as the crest

wave refraction
the modification of both the wave direction and the wave height by depth variations for intermediate and shallow-water waves propagating shoreward exclusive of wave shoaling

wave shoaling
increase in a wave's height caused by the decreasing depth as it travels into shallower water

wave train
a group of waves

wave trough
the distance from the undisturved water surface to the minimum wave amplitude

BIBLIOGRAPHY

Atwater, B. F., S. Musumi-Rokkaku, K. Satake, Y. Tsuji, K. Ueda, and D. K. Yamaguchi. 2005. The Orphan Tsunami of 1700—Japanese Clues to a Parent Earthquake in North America. Professional Paper 1707, U.S. Geological Survey.

Bird, E. C. F. 1996. *Beach Management*. Wiley.

Bretschneider, C. L. 1952. Revised Wave Forecasting Relationships. In Proceedings of Second Conference on Coastal Engineering, Berkeley.

Dean, R. G. 1973. Heuristic Models of Sand Transport in the Surf Zone. In Proceedings of Conference on Engineering Dynamics in the Surf Zone, Sydney.

Dean, R. G. 1991. Equilibrium Beach Profiles: Characteristics and Applications. *Journal of Coastal Research* 7 (1): 53–84.

Dean, R. G. 2002. *Beach Nourishment: Theory and Practice*. World Scientific.

Dean, R. G., and R. A. Dalrymple. 1984. *Water Wave Mechanics for Engineers and Scientists*. Prentice-Hall.

Dean, R. G., and R. A. Dalrymple. 2002. *Coastal Processes with Engineering Applications*. Cambridge University Press.

Dronkers, J. J. 1964. *Tidal Computations in Rivers and Coastal Waters*. North-Holland.

Goda, Y. 1985. *Random Seas and the Design of Maritime Structure*. University of Tokyo Press.

Gordon, A., W. Grace, P. Schwerdtfeger, and R. Byron-Scott. 1998. *Dynamic Meteorology: A Basic Course*. Wiley.

Hammack, J. L. 1973. A Note on Tsunamis: Their Generation and Propagation in an Ocean of Uniform Depth. *Journal of Fluid Mechanics* 60 (4): 769–799.

Hanks, T. C., and H. Kanamori. 1979. A Moment Magnitude Scale. *Journal of Geophysical Research* 84 (B5): 2348–2350.

Hasselmann, K. 1961. On the Nonlinear Energy Transfer in a Wave Spectrum. In *Ocean Wave Spectra*. Prentice-Hall.

Hasselmann, K. 1962. On the Nonlinear Energy Transfer in a Gravity Wave Spectrum, Part 1, General Theory. *Journal of Fluid Mechanics* 12 (4): 481–500.

Heaton, T. H., and P. D. Snaveley. 1985. Possible Tsunami along the Northwestern Coast of the United States Inferred from Indian Traditions. *Bulletin of the Seismological Society of America* 75 (5): 1455–1460.

Helmholtz, H. 1888. *Sitz*. Akademie der Wissenschaften (Berlin).

Helmholtz, H. 1890. *Die Energie der Wogen und des Windes*. Akademie der Wissenschaften (Berlin).

Herbich, J. B. 1990. *Handbook of Coastal and Ocean Engineering*, volume 1. Gulf.

Houston, J. R. 1996. International Tourism and U.S. Beaches. *Shore and Beach* 64 (2): 3–4.

Hughes, S. A. 1993. *Physical Models and Laboratory Techniques in Coastal Engineering*. World Scientific.

Iida, K. 1961. Magnitude, Energy, and Generation Mechanisms of Tsunamis and Catalogue of Earthquakes Associated with Tsunamis. In Proceedings of the Tsunami Meetings Associated with the Tenth Pacific Science Congress, Hawaii.

Ippen, A. T. 1966. *Estuary and Coastline Hydrodynamics*. McGraw-Hill.

Jeffreys, H. 1924. On the Formation of Water Waves by the Wind. *Proceedings of the Royal Society of London Series A* 107 (742): 189–206.

Komar, P. D. 1998. *Beach Processes and Sedimentation*, second edition. Prentice-Hall.

Kraus, N. C. 1996. *History and Heritage of Coastal Engineering*. American Society of Civil Engineers.

Lee, J. J., X. Xing, and O. T. Magoon. 2008. Uncovering the Basin Response at Crescent City Harbor Region. In Proceedings of 31st International Conference on Coastal Engineering.

Lillevang, O. J., F. Raichlen, J. C. Cox, and D. L. Behnke. 1984. A Detailed Model Study of Damage to a Large Breakwater and Model Verification of Concepts for Repair and Upgraded Strength. In Proceedings of 19th International Conference on Coastal Engineering, Houston.

McCarthy, R. J., E. N. Bernard, and M. R. Legg. 1993. The Cape Mendocino Earthquake: A local tsunami wakeup call? In *Coastal Zone '93: Proceedings of the Eighth Symposium on Coastal and Ocean Management, New Orleans*. American Society of Civil Engineers.

Neiburger, M., J. G. Edinger, and W. D. Bonner. 1982. *Understanding Our Atmospheric Environment*. Freeman.

Neumann, G., and W. J. Pierson, Jr. 1966. *Principles of Physical Oceanography*. Prentice-Hall.

Ochi, M. K. 1982. Stochastic Analysis and Probabilistic Prediction of Random Seas. *Advances in Hydroscience* 13: 218–375.

Phillips, O. M. 1957. On the Generation of Waves by Turbulent Wind. *Journal of Fluid Mechanics* 2 (5): 417–445.

Pierson, W. J., Jr., G. Neumann, and R.W. James. 1955. Practical Methods for Observing and Forecasting Ocean Waves by means of Wave Spectra and Statistics. Publication 603, U.S. Navy Hydrographic Office (reprinted 1960).

Poon, Y.-K., F. Raichlen, and J. Walker. 1998. Application of a Physical Model in Long Wave Studies for the Port of Long Beach. In Proceedings of 26th International Conference on Coastal Engineering.

Raichlen, F. 1975. The Effect of Waves on Rubble-Mound Structures. *Annual Review of Fluid Mechanics* 7: 327–356.

Sarpkaya, T. and M. Isaacson 1981. *Mechanics of Wave Forces on Offshore Structures*. van Nostrand Reinhold.

Savile, L. H. 1940. Presidential Address on Ancient Harbor. *Journal of the Institution of Civil Engineers* 15: 1–26.

Sorensen, R. M. 1993. *Basic Wave Mechanics: For Coastal and Ocean Engineers*. Wiley.

Spaeth, M. G., and S. C. Berkman. 1967. The tsunami of March 28, 1964, as Recorded at Tide Stations. Technical Bulletin 33, U.S. Coast and Geodetic Survey, Rockville, Maryland.

Sverdrup, H. U., and W. H. Munk, 1947. Wind, Sea and Swell: Theory of Relations for Forecasting. Publication 601, Hydrographic Office, Department of the Navy.

Thomson, W. 1887. On the Waves Produced by a Single Impulse in Water of Any Depth, or in a Dispersive Medium. *Proceedings of the Royal Society of London* 42: 80–83.

U.S. Army Corps of Engineers. 1984. *Shore Protection Manual*.

U.S. Army Corps of Engineers. 2001. *Coastal Engineering Manual*.

Wells, N. 1997. *The Atmosphere and Oceans*. Wiley.

Wiegel, R. L. 1964. *Oceanographical Engineering*. Prentice-Hall.

SUGGESTED READING

Bascom, W. 1964. *Waves and Beaches*. Doubleday.

Bird, E. C. F. 1996. *Beach Management*. Wiley.

Butt, T., and P. Russell, with R. Grigg. 2002. *Surf Science: An Introduction to Waves for Surfing*. University of Hawaii Press.

Casey, S. 2010. *The Wave: In Pursuit of the Rogues, Freaks, and Giants of the Ocean*. Doubleday.

Dean, S. 2010. *The Wave: In Pursuit of the Rogues, Freaks, and Giants of the Ocean*. Doubleday.

Dean, R. G., and R. A. Dalrymple. 1984. *Water Wave Mechanics for Engineers and Scientists*. Prentice-Hall.

Ippen, A. T. 1966. *Estuary and Coastline Hydrodynamics*. McGraw-Hill.

Komar, P. D. 1998. *Beach Processes and Sedimentation*, second edition. Prentice-Hall.

Neiburger, M., J. G. Edinger, and W. D. Bonner. 1982. *Understanding Our Atmospheric Environment*. Freeman.

Sorensen, R. M. 1993. *Basic Wave Mechanics for Coastal and Ocean Engineers*. Wiley.

U.S. Army Corps of Engineers. 2001. *Coastal Engineering Manual*.

Wiegel, R. L. 1964. *Oceanographical Engineering*. Prentice-Hall.

INDEX

Printed in the United States
by Baker & Taylor Publisher Services